Beginner's Guide to

Electric Wiring

F Guillou
Lecturer in Electrical Engineering,
Old Swan Technical College, Liverpool

C Gray, BSc
Lecturer in Electrical Engineering,
Old Swan Technical College, Liverpool

Newnes Technical Books

Newnes Technical Books
is an imprint of the Butterworth Group
which has principal offices in
London, Boston, Durban, Singapore, Sydney, Toronto, Wellington

First published 1965
Second edition 1975
Reprinted 1976, 1977, 1979, 1981
Third edition 1982
Reprinted (with revisions) 1983

British Library Cataloguing in Publication Data

Guillou, F.
Beginner's guide to electric wiring – 3rd ed.
1. Electric wiring, Interior
I. Title II. Gray, C.
621.319'24 TK3271

ISBN 0-408-01130-0

Printed in England by Whitstable Litho Ltd, Whitstable, Kent

Preface

This book is intended to bridge the gap between the layman and electrical tradesman, with regard to electric wiring in homes and workshops. Although the work is not intended as a textbook for students, the information given complies with the 15th Edition of the *Regulations for Electrical Installations* published by the Institution of Electrical Engineers, and much of the content will be useful to first year electrical students in the electrical contracting industry. The reader need not have any technical knowledge at the start. Only simple calculations are involved.

The authors emphasise the need for adequate and safe electric wiring in the home, as too often death and injury result from unsafe electrical installations. It is hoped that this book will reduce the number of dangerous situations arising out of the layman's mistaken understanding of the correct methods of electric wiring.

We would like to thank the IEE for permission to publish extracts from the *Regulations for Electrical Installations*, 15th edition.

F.G.
C.G.

Contents

1

Introduction

The purpose of electric wiring is to make available electrical energy whenever it is required:

(1) With maximum safety.
(2) With the capability of supplying the current for the usage required and for possible future extended usage.
(3) With maximum reliability.
(4) With maximum flexibility to provide for change in usage and extension.
(5) Economically.

Terminology

Terms used throughout this book include the following:

Wiring
The fixed installation of insulated electric cables between the intake point in a particular installation and the appliances that use the current (such as lamps, radiators, cookers, radio sets, etc.), including the fuses, switches, socket-outlets, lampholders and all other parts permanently fixed in the building.

Appliances
All current-consuming apparatus, whether fixed (such as a plumbed-in water heater), or portable (such as an electric iron).

Socket-outlet

A properly designed and permanently fixed device installed so that a portable appliance can be safely plugged in. (In Britain, although not in some other countries, all socket-outlets should be of the three-pin type, with two pins for the circuit connections and one for the earth connection.)

Area electricity board

In Britain, all electricity supplies are given by one or other of the twelve area electricity boards (in England and Wales) and equivalent authorities in Scotland. For the purposes of this book, the term 'area board' may be taken as referring to the electricity supply authority, whether it is a board, a power company, or any other body.

Series connection

A form of connection (*Figure* 1.1) in which all the current passes through the circuits or appliances one after the other.

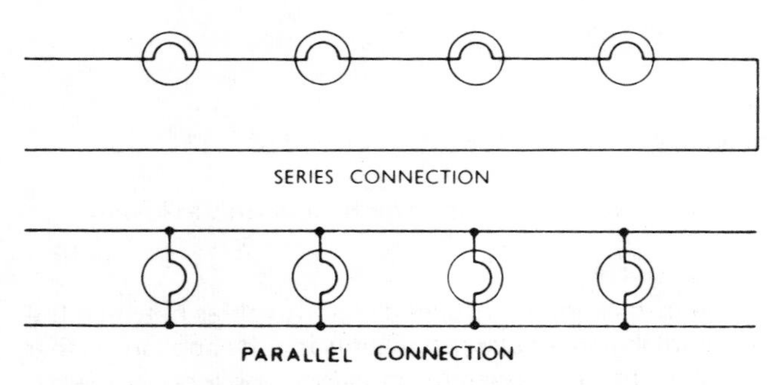

Figure 1.1. Series connection and parallel connection

Parallel connection

A form of connection in which all current-consuming parts of the circuit are individually connected across the two wires providing the supply.

Power supply

The source of power

Electricity is generated in power stations where coal or oil is burnt to produce steam which turns a turbine coupled to an electrical generator. In Britain a number of nuclear power stations also provide power, the heat from the atomic reaction serving the same purpose as burning fuel in the furnace. In Britain there is a small proportion of power generated by falling water, driving a water wheel, or water turbine, but in some countries the majority of the power supply comes from this source.

The power is almost universally generated as alternating current, where the direction of flow of current changes fifty times a second (50 Hz (hertz)). In America and in some other parts of the world, the frequency is 60 Hz. Alternating current is used in preference to direct current mainly for ease of transforming from high to low voltage, and vice versa.

Obviously, the more current that is needed the larger the conductor necessary at a given pressure (voltage). This can be understood by reference to a water pipe system. To fill a given tank in a given time, either a high-pressure hose, of small diameter, can be employed, or else a low pressure and a large diameter hose. It is the same with electricity, where the pressure is represented by the voltage, and the flow by the current. To carry power from a power station to the point of consumption, perhaps fifty or one hundred miles away, overhead grid systems are used. It is impossible to increase the size of electrical conductor carried on pylons beyond a certain practical limit. The only way, then, to carry more power, is to increase the pressure, or voltage.

By the use of the transformer, which will be mentioned later in this book, alternating current can be transformed up or down in voltage, as required, and the voltage used for bulk transmission on the grid system is as high as 750,000 V.

This voltage is transformed down until it reaches the transformer at the end of one's own street, or somewhere on an industrial or housing estate site, at a voltage of 11,000 V; and the final transformation, to feed the power into the cables connected to the consumer's premises, is to a voltage of 415/240 V, 3-phase.

All alternating current power is generated on the 3-phase system. This means that the part of the generator where the rotating magnetic field sets up the current we ultimately use is divided into three equal sectors. The three separate windings, in which the power is generated, are brought out by means of six wires, one at the end of each winding, and one end of each winding has the wires connected to a common point, known as the neutral. The other ends – the free ends – are the supply mains, usually known as phases, and for ease of identification are called the red, yellow and blue phases.

The 3-phase system continues all the way to the transformer near to the consumer's premises, which we will call the local transformer, and on the low voltage side of this transformer we thus have four wires, the red, yellow and blue phases and the neutral. These wires are identified with red, yellow and blue markings, with black for neutral (*Figure* 1.2).

In the 3-phase system, two voltages exist. Between each phase and the neutral wire, on the low voltage side of the local transformer, the voltage is 240 V, the standard voltage in Britain (in other countries this may differ, and in America, for example, 110 V is commonly used). But between the red and yellow phases there is a voltage of 415 V, and 415 V also exists between the yellow and blue, and between the blue and red. This is the reason why the output voltage of the local transformer is given as 415/240 V.

The cables running out from the local transformer to the premises of the consumer (usually underground, but in rural districts overhead, on wooden poles) are tapped off, for each house, by taking a connection, say for the first house in the street from the red phase wire in the 3-phase cable and from the neutral, thus giving a 240 V supply to that house; for the second house, from the yellow phase wire and the neutral; and for the third house from the blue phase wire and the neutral.

This form of connection is made in order to balance the demand on the three phases, since the generator must

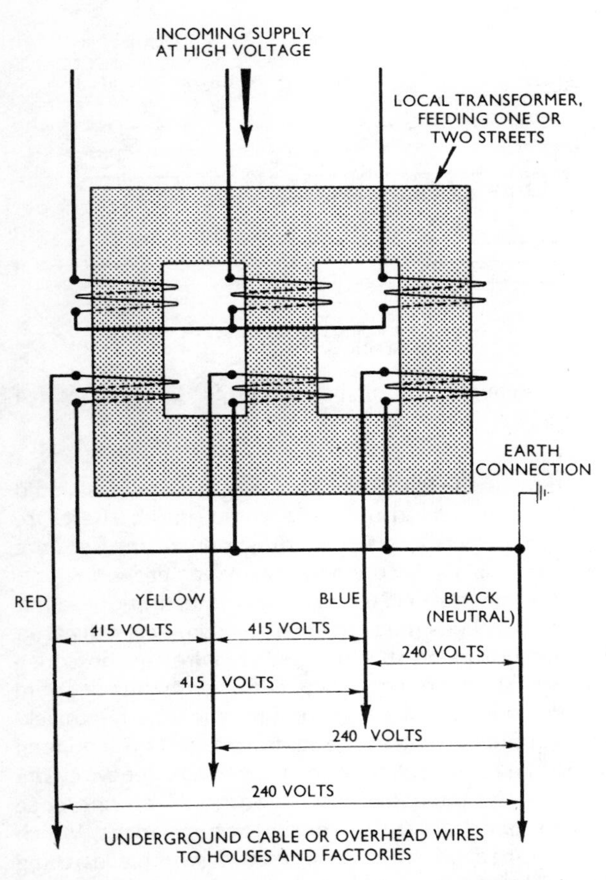

Figure 1.2. The transformer steps down the high voltage supply from the power station and provides a 3-phase supply

ultimately supply a balanced load, and the varying demands of the thousands of consumers that might be connected to any one generator will balance out if connections are made in this way *Figure* 1.3).

However, it is usual, where the demand in a particular consumer's premises exceeds 15 kW, for a 3-phase supply to

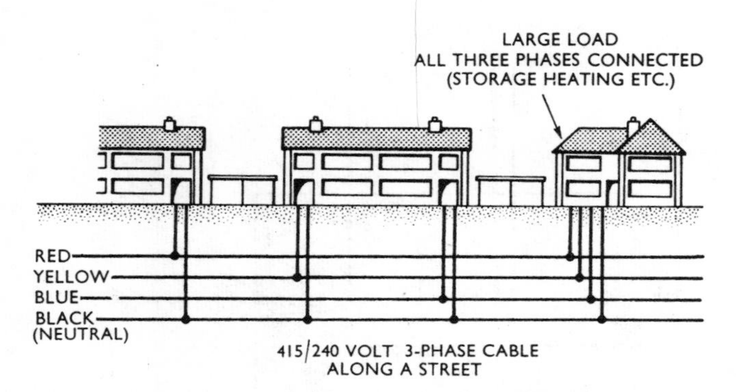

Figure 1.3. System of balancing the loads on the three phases in a suburban street

be given, because the load would be too great to be balanced out properly if connected only to a single phase. Therefore more and more consumers, including private houses, are being fed with a 3-phase supply, as the loads grow.

At the local transformer, the neutral connection of the low voltage side, that feeds the consumers, is connected to earth. But this does not mean that the neutral wire can be considered as safe. There are circumstances under which it could become live, and although in general the neutral (black) connection appears to be at the same voltage as the general mass of earth, it must at all times be treated as a live wire. The neutral wire is *not* the earth connection, to which reference will be made later, except in special circumstances, which will be mentioned later under protective multiple earthing (see p. 65). The definition of 'live' includes the phase and neutral conductors.

The circuit

It is fundamental to electric current that there must be a circuit (see *Figure* 1.4). This means that the current must be able to flow from the point where it originates, at the local

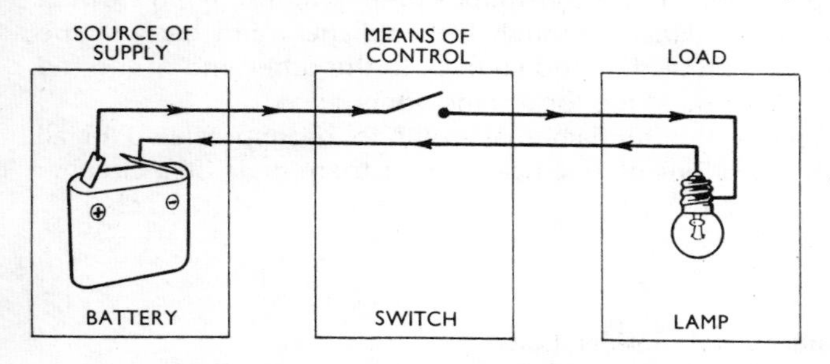

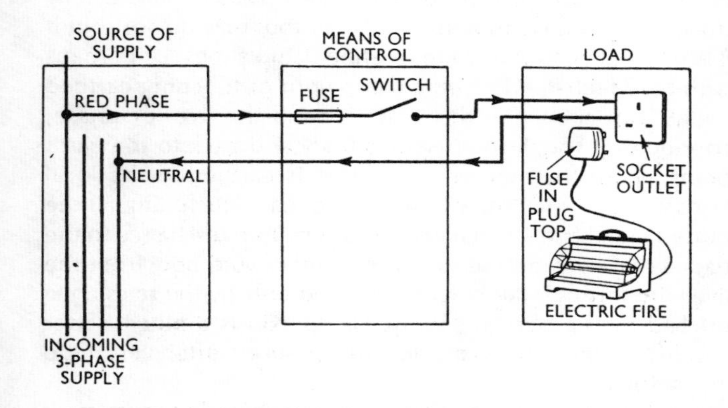

Figure 1.4. The principle of the electric circuit: *above*, the simplest electrical circuit; *below*, the normal alternating current circuit used on domestic premises (Protective (earth) conductor not shown)

transformer, on, say, the red phase terminal, through the supply cable to the consumer's premises, through his wiring to the appliance, through the appliance, and back to the neutral connection and so back to the other end of the red phase wiring at the local transformer.

This is the fundamental point to be appreciated in all considerations of electrical wiring; there must be a circuit.

Fundamental considerations

Safety

Electricity is a good servant but a dangerous master. The lowest recorded voltage at which death occurred from an electric shock is 38V. In general, 240V seldom kills a fully dressed person wearing dry footwear (but may do so): but it is a fatal voltage for anyone wearing damp shoes, or perhaps with bare feet standing on a damp floor or touching earthed metal. The circuit, in this case, is from the live, or phase, conductor through the person's body and back to the earth point where the neutral connection is earthed in the local transformer. Since the whole mass of earth including all the buried metal and so on usually has very little resistance to the passage of current, the full current that could flow from the live wire through the person is limited only by the resistance offered by the person's body. Damp skin is a much better conductor than dry skin, and damp shoes offer very little insulation.

Anyone attempting to carry out electric wiring must at all times remember that he is dealing with a potentially lethal form of energy.

In Britain, it is still permissible for anyone to extend his or her electric wiring system, or to install new wiring, without special qualifications. In many countries this is not the case. It is a punishable offence, for example, in New Zealand, for anyone other than a registered and qualified electrician to install wiring of any kind.

However, if the work is undertaken with a full sense of responsibility, proper materials are used and proper methods employed, there is no reason why a safe installation should not result. *But the person installing wiring should always have in the front of his mind the possibility that he might have to give evidence at a coroner's court.*

The safety of electrical appliances and wiring is ensured, basically, in three ways. First, by *insulation*; secondly, by *earthing*; and thirdly, *by proper protection against fire risk.*

Insulation

Insulation is the method whereby the live electric wires or other equipment are covered in such a way that it is impossible for anyone to come into contact with live metal. Wiring, for example, is made up of copper conductors (sometimes aluminium conductors) covered with insulating material, and if the right type of cable is used, there can be no danger from touching the outside of the insulating covering.

Appliances of all kinds, if of proper design, have the live parts completely encased in porcelain or plastic materials so that it is impossible even for an inquiring child to insert its finger into any part that is live. Otherwise the live equipment is fitted inside a sealed part of the appliance, and access can only be obtained to it by deliberate interference.

Double insulation

Certain appliances are of what is known as the 'double-insulated' type. These appliances have first the normal functional insulation, as in all other appliances, and then a separate protective insulation enclosing all metal parts. Such appliances do not need an earth connection, but it must be borne in mind that no appliance can be considered as double insulated unless it complies with the Regulations and has been certified by the British Electrical Approvals Board.

Some (but not all) designs of shavers, hair dryers, dishwashers, clocks, blankets and similar appliances have been certified as double insulated, and thus need only the phase and the neutral connections to the mains.

Earthing
Earthing is the second line of defence. The whole mass of earth is obviously safe from the electrical point of view. Therefore if, say, a kettle has a wire connecting the body of the kettle, which can be touched, to the earth, then whatever happens to the live wires inside the kettle heating element,

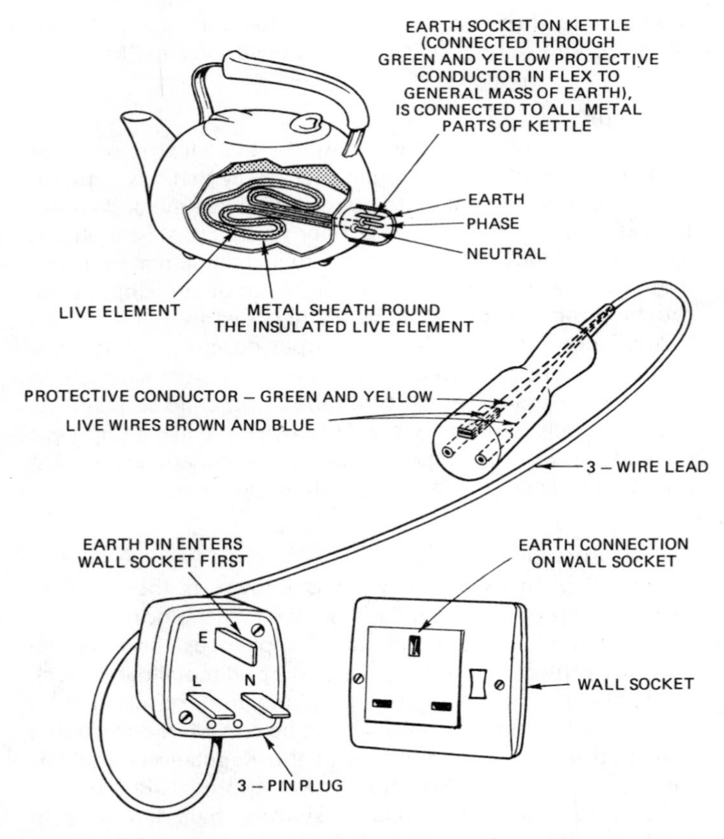

Figure 1.5. The earthing of a portable appliance, showing the connection through the 3-wire flexible lead to the earth point on the socket-outlet

or in the connector feeding the appliance with current, the user is safe because any current finding its way to the body of the appliance would be short-circuited straight to earth (see *Figure* 1.5).

Metal parts of any kind of appliance in which electricity is used should always be earthed (with the exception of properly certified double-insulated appliances). With portable appliances, this is carried out, as will be seen later, by using a 3-core flexible cable and a proper plug and socket-outlet, the third pin of which, the earth pin (separate and different from phase and neutral pins) is properly connected to the earth. This is done by connecting the earth terminal in the socket-outlet on the wall to a proper earth point, which (by arrangement with the area electricity board only) may be earthed to the lead sheathing of the supply cables at the consumer's terminal point, or may be a special earth prepared properly and provided for this purpose.

At the appliance end, the connector or the cable termination (if it is a fixed termination) must be so arranged that every metal part of the appliance is properly connected to a terminal to which the third wire in the flexible – green and yellow – is connected, so that whatever leakage of current might take place, in whatever part of the appliance, the current will flow harmlessly to the earth point, through the green and yellow insulated protective conductor into the plug and to earth via the proper earth point.

Protection against fire risk

Protection against fire risk is secured by using properly dimensioned cables and fittings, and by protecting the circuits by means of the correct sizes and types of fuses or circuit-breakers.

Properly designed wiring

As well as the danger of shock, the supply of electric power from a large power station brings with it another danger.

When current passes through a wire, there is a certain

resistance in the wire to be overcome. In overcoming this resistance, heat is generated in the wire. In the ordinary open-type electric radiator, this heating effect is usefully employed to give the heat we require from the radiator.

But heat is also being generated in the wiring itself. If the wiring is properly sized, the heating effect is small, but if the wire is too small, two problems arise. First, in extreme cases there is danger of fire from the wiring becoming too hot, and setting fire to adjacent materials with subsequent damage to the cable insulation, and secondly some of the voltage in the supply mains will be lost in the cable itself, and the appliance will not receive its full voltage, and thus, for example, lights may be dimmer than they should be.

This question of the proper sizing of the wiring installation in relation to the load to be fed with current applies not only to the wiring itself but to all the appliances used. Even the ordinary switch found on the wall of a living room must be suitable for its duty. Poorly designed switches, or switches that have become worn out, cannot carry the current they were intended to carry, or the increased current that wiring extensions have made possible for them to carry, mainly because the contact parts within the switch were never large enough, or have become twisted or bent or overheated so that they no longer make good contact. In consequence, there is too high a resistance within the switch, further heat is generated, and fire may result. This factor also applies to fuses, to junction boxes and to any part of the installation such as lampholders. There is a special aspect of the fire risk to be borne in mind when lampholders are considered.

People are becoming accustomed to higher and higher levels of lighting than were accepted in the past, and they have tended to add larger and larger lamps to existing lampholders. In addition, mushroom-shaped lamps are now available which give a high wattage, or power consumption, in a small volume, and lighting fittings are being made to accommodate lamps of a greater power than those for which they were designed.

A lamp gives out almost all its power in the form of heat. The overheating which may take place when oversized lamps are used may give rise to serious consequences, because the flexible or other wiring connected to the lampholder may become overheated to the stage where its insulation is damaged, and a short circuit may occur, which at the least may black out a number of lights, or at worst give rise to a fire.

Protection of the installation

The most commonly used type of protection is the fuse (*Figure* 1.6). A fuse is like a weak link in a chain, carefully designed to break when the maximum permissible load is exceeded, so that the crane, for example, to which a chain might be connected cannot be overloaded and perhaps overturned.

In the electric circuit, a fuse consists usually of a fine wire which has been carefully selected so that, when the maximum current which the circuit should carry is exceeded, the wire melts and effectively switches off the current, so saving the wiring itself and all the appliances on the circuit from damage.

Fuses are usually contained in fuseholders and installed in a consumer unit, in such a way that the designed overheating of the fuse and its ultimate rupture by melting cannot give rise to any fire risk or other serious consequences.

The circuits in the installation must be fused according to the Regulations mentioned later, and care must be taken to see that a blown fuse is replaced with a fuse of the proper type for the duty required, and not with a larger type which would invalidate the protection it gives to the circuit.

It is extremely unwise to repair a fuse that has blown without finding out why it blew. It might well happen that the danger still exists, for example if a flexible cord is frayed through it may have shorted its two conductors together sufficiently to have blown the fuse, and then left the bare copper exposed, so that a shock could result if the fuse is replaced. The faulty part of the installation should either be repaired or temporarily isolated, before replacing the fuse that has blown.

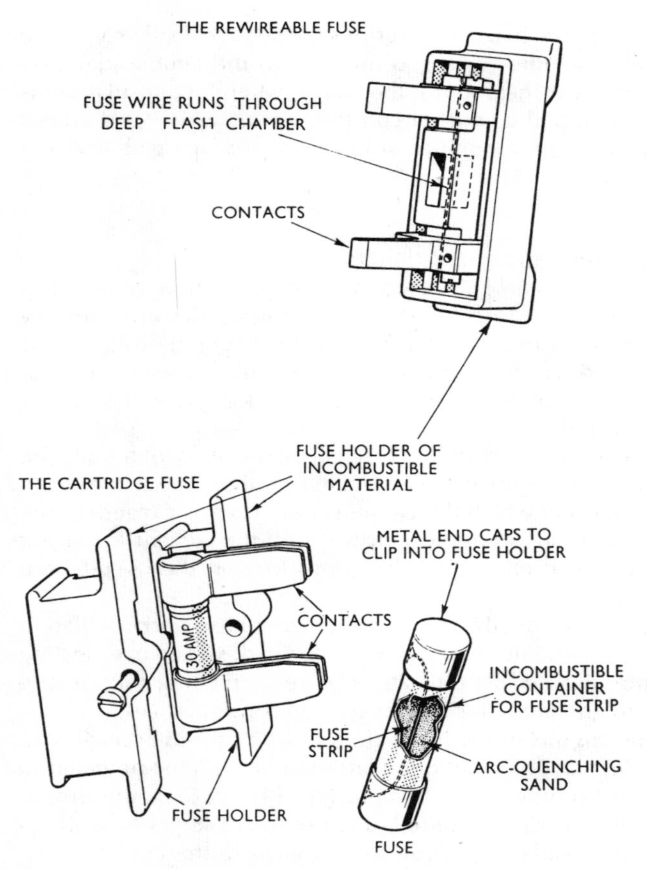

Figure 1.6. Two types of fuse: *above*, the rewireable fuse; *below*, the cartridge fuse

The miniature circuit breaker, or MCB as it is often referred to, may be used to protect a circuit. The MCB is a switch which operates automatically in the event of an overload or short circuit. It has the advantage that once installed, with the correct current rating for the circuit, its rating cannot be changed.

Reliability
Since most of us rely entirely on our electrical systems, anyone installing any kind of wiring must be careful to ensure that the wiring is absolutely reliable. This means that not only must it be properly designed but that it must be laid out in the building in such a way that it is not likely to be subjected to casual damage, and that any failure of wiring, or of appliances connected to any particular circuit, must not be allowed to give rise to wholesale blackouts and failures of supply in other parts of the building.

Flexibility
Wiring requirements are constantly changing. With the addition of more and more domestic appliances, and more and more electrical services in offices (for lighting, office machinery and every kind of electronic aid to office working), while in small factories and workshops additional electrical appliances are constantly being added, thought should be given to planning the installation so that it does not become overloaded. This may be effected by providing for growth of the installation by allocation of additional fuses etc.

Cost
There are many people who naturally feel that wiring must always be carried out as cheaply as possible. It is safe to say that good wiring pays for itself in peace of mind and in ease and convenience of usage. Money skimped on a wiring installation is money unwisely skimped. The minimum cost consideration should never be in the forefront of the mind of anyone planning or executing a wiring installation.

The electrician's responsibility

In every installation, the area electricity board or the electricity supply company or authority provides a service terminal point, on which all parts are sealed (*Figure* 1.7). This usually consists of a main fuse, or cut-out, and a meter. Some

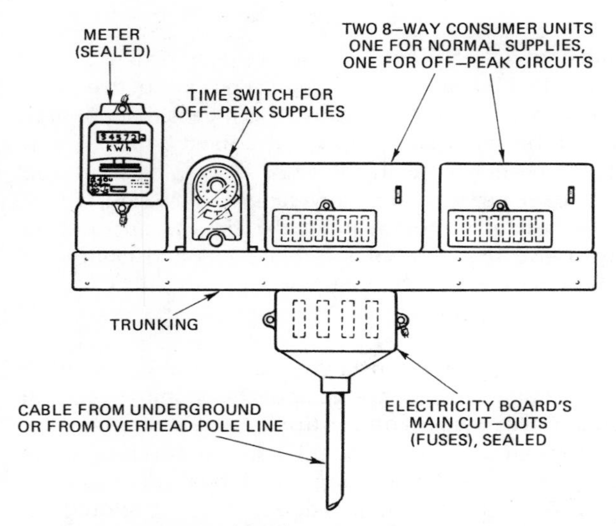

Figure 1.7. Consumer's terminal showing trunking to carry wiring to meter, consumer units and time switch

consumers take advantage of 'off-peak' current and, as far as possible, restrict their water heating and space heating loads to the off-peak period, i.e. during the night. In such cases the electricity board provide a time switch to control the off-peak usage, which is about half the cost of normal usage. A dual-purpose meter is also fitted by the electricity board which records separately the normal and off-peak consumption.

All these items are the property of the board, and must not be interfered with in any way by the electrician. To break the seal on the main fuses or any part of the metering circuit is to invite prosecution.

After the meter, the responsibility for the installation lies entirely with the consumer, and if it is a new installation, the electrician is therefore responsible for handing over to the consumer a proper wiring installation, which the electricity board will test, and if found in good order will connect up to the meter.

The electrician's bible

The main guidelines for electric wiring are the Regulations of the Institution of Electrical Engineers, known as the *Regulations for Electrical Installations*. These Regulations are amended from time to time; so the user must make certain that he has the latest copy.

These Regulations, although they are not part of the law of the land, are very nearly in the same category. For example, an insurance company, if it is asked to insure a building, will nearly always specify that the electric wiring must be in conformity with the practice laid down in the Regulations of the Institution. In cases where accidents occur, such as fires or electric shocks, the best possible defence, on the part of the person who installed the wiring, is that it is in conformity with the Regulations. One would have a very poor defence indeed if the wiring did not conform to the Regulations.

Tests by electricity board

The electricity board has the right to refuse to connect up an installation which it considers to be unsafe, and such an installation obviously does not comply with the Regulations. The tests the board's engineers will apply will be directed to ensuring that the Regulations have been properly carried out.

The electricity board also has the right to be notified when any alteration or addition is made to an installation, and may inspect the altered or added parts before they are connected to the main system.

Some definitions

Voltage

The pressure that forces the current round an electric circuit is measured in volts (V). The normal domestic supply (on systems with standard supplies) is at 240V: a flashlamp bulb works at 1.5V: most of the National Grid System works at

275,000 V, parts at 400,000 V. The voltage is equivalent to the head of water causing a flow along a pipe.

The following voltage levels are defined:

Extra-low voltage: below 50 V (120 V for d.c.) between conductors or to earth.

Low voltage: exceeding extra-low voltage but not exceeding 1000 V (1500 V for d.c.) between conductors, or 600 V (900 V d.c.) between any conductor and earth.

The electrician, dealing with domestic and small industrial installations, will therefore deal mainly witl. low-voltage supplies.

Current

The flow, or current, in a conductor is measured in amperes (A). A 1 kW fire needs a flow of about 4.16 A: a 100 W lamp needs about 0.4 A.

Resistance

When water flows through a thin pipe, it encounters a resistance to its flow. The larger the pipe, the less resistance. Also pipes differ, in resistance to flow, even at the same diameter. A smooth-bore pipe will offer less resistance than a rusted bore.

Similarly, with conductors of electricity, the thicker the wire, in general the less the resistance. Certain wires, like copper, silver and aluminium, have less resistance than similar-sized wires of steel or nickel alloy.

The effect of resistance

The effect of resistance in a conductor is to generate heat as the current passes, as mentioned earlier. In radiator elements, this is the desired effect, and the conductors used in these elements are designed to give a suitable resistance to generate the required amount of heat.

But in the conductors used for wiring a house, the aim should always be to reduce this heating effect to the minimum. Heat is not required in the wiring system: it will reduce the life of the insulation, and if excessive may even cause a fire.

Therefore care must always be taken to reduce the heating effect to the minimum by ensuring that all conductors used – the cables themselves, and all the fittings of every kind – are large enough to present a very low resistance to the current for which the circuit is to be used.

Resistance is measured in ohms (Ω). As an example, a 100 W lamp has a resistance of about 600 Ω (though this varies a little as it heats up).

The resistance of the insulation used on electrical appliances is of course very high indeed – otherwise it would not be regarded as insulation. To give a typical example, the resistance of the insulation used in a good, new electric iron, measured between the live conductors in the element and the outer metal case of the iron, ought to be approximately 2,000,000 Ω known as 2 *megohms* (MΩ).

The resistance of all parts of an electrical installation in a domestic dwelling – that is, the resistance of all the live conductors in the cables, the switches, the socket-outlets, and the consumer unit – measured against earth, must not be less than 1,000,000 Ω.

Later we shall see how this insulation resistance is tested.

Two simple formulae

There are two very simple and fundamental formulae that must be understood in relation to all electric circuits.

The ability of an appliance of any kind to consume electricity is measured in watts (W). A thousand watts equals one *kilowatt* (kW).

An appliance that is so designed that its resistance allows 1 A of current to flow when the appliance is connected to a voltage of 1 V is said to have a power rating of 1 W.

Therefore

$$\text{power (W)} = \text{current (A)} \times \text{voltage (V)}$$

As an example, a one-bar radiator with an element of 1000 W (1 kW) rating would take current of 4.16 A if connected to a 240 V circuit:

$$\text{power (W)} = \text{current (A)} \times \text{voltage (V)}$$
$$1000 = 4.16 \times 240$$

If you know any two of these three figures, you can calculate the other.

Often it is desired to know what current will be needed to feed a certain appliance, say a television set with a nameplate that says 200 W. We proceed as follows, assuming the set is to be connected to the 240 V supply mains:

$$\text{power (W)} = \text{current (A)} \times \text{voltage (V)}$$

Therefore:

$$\frac{\text{power (W)}}{\text{voltage (V)}} = \text{current (A)}$$

so,

$$\frac{200}{240} = \tfrac{5}{6} \text{ of an ampere, or } 0.83\,\text{A}$$

Ohm's law

The relation between the voltage and the current, in a simple circuit that has a certain resistance, is as follows:

$$\text{current (A)} = \frac{\text{voltage (V)}}{\text{resistance } (\Omega)}$$

To give a simple example, a kettle of 3 kW rating will have a resistance (in its heating element) calculated as follows:

$$\text{power (W)} = \text{current (A)} \times \text{voltage (V)}$$

Therefore:

$$\frac{\text{power (W)}}{\text{voltage (V)}} = \text{current (A)}$$

$$\frac{3000}{240} = 12.5\,\text{A}$$

From Ohm's law:

$$12.5\,\text{A} = \frac{\text{voltage (V) (240)}}{\text{resistance } (\Omega)}$$

Therefore:

$$\text{resistance } (\Omega) = \frac{240}{12.5} = 19.2\,\Omega$$

Note: These simple but fundamental formulae apply to direct current and ordinary alternating current circuits in which the appliances used – radiators, immersion heaters, filament lamps, and the like – are mainly users of electricity by means of resistance wires that grow hot when current flows. When appliances that use motors and such devices as the electromagnetic choke coils in fluorescent lamps are considered, the formulae have to be modified to take into account the electromagnetic effects of the current: but the modifications needed for domestic and similar installations are usually so small that they can be ignored. They only become significant where larger industrial installations are concerned.

The unit of electricity

Although not strictly a matter concerning the electrical installation itself, it is as well to complete the picture by mentioning the current usage of various appliances.

An appliance – say, a one-bar electric fire – is designed so that when it is connected to the usual 240 V mains, it will use 1000 W, that is 1 kW.

If such a 1 kW fire is switched on for exactly 1 hour, it will draw 1 kilowatt-hour of energy from the mains.

One kilowatt-hour is the standard unit of electricity consumption.

You can use one unit of electricity by:

burning a 1 kW fire for 1 hour,
burning a ½ kW fire for 2 hours,
burning a 2 kW fire for ½ hour.

To take another example, a 25 W lamp used for 1 hour:

$$25\,W \times 1 \text{ hour} = 25\,Wh = \tfrac{1}{40}\text{th of } 1000\,Wh = \tfrac{1}{40}\text{th of a unit}$$

or,

$$25\,W \text{ for } 40 \text{ hours} = 25 \times 40\,Wh = 1000\,Wh = 1 \text{ unit.}$$

One further example: a cooker hotplate has a capacity, or loading, of 2 kW.

$$2\,kW\ (=2000\,W) \text{ for } 1 \text{ hour} = 2\,kW = 2 \text{ units.}$$

2

Materials for wiring installations

Wiring: general

Wiring may be carried out in several ways, which are conveniently discussed under four headings:

the conductor
the insulation
the sheath
the enclosure.

The conductor

In the majority of cases the conductors used in domestic or small commercial wiring installations are made of high conductivity copper, sometimes covered with a coating of tin to assist in preventing corrosion. In some cases, however, aluminium conductors are used.

Often, a single strand of conductor is used but to give more flexibility to heavier duty cables a number of small wires are stranded (twisted) together to make a larger conductor.

The important aspect of the conductor is of course the total cross-sectional area of the metal, since the larger the metal area, the greater the current-carrying capacity (subject to certain other factors). For example, a wire size very commonly used is known as $2.5\,mm^2$. This means that the conductor has a cross-section area of $2.5\,mm^2$.

The insulation

To insulate the live conductors, various methods are used.

Plastic insulation

There are several forms of plastic materials that are commonly used to insulate wiring conductors. First there is polyvinyl chloride (PVC) which is extremely tough, will not support combustion, and is not affected by water, oil, or most chemicals. It has however a definite temperature limit, above which it will melt and leave the conductor bare.

Then there is polythene insulation, made from a different type of plastic, with a lower safe temperature rating.

PVC/SWA cable uses PVC insulation, but is enclosed in a protective wrapping of steel wires, to form an 'armouring' or protective covering. This type of cable is normally used where a high degree of mechanical protection is required, such as underground between buildings.

Rubber insulation

This is the oldest form of insulation, and is still used, but is largely superseded by plastic materials for all but special applications.

Rubber is very flexible, but has a number of disadvantages. It burns easily, and it can be attacked by chemicals and by oil. It is to some extent absorbent to water. Direct sunlight soon causes rubber to deteriorate, so unprotected rubber cables should never be used outdoors. It is also attacked by certain insects, and in any case it ages and tends to become brittle.

The sheath

The insulated conductor is more often than not grouped with others within an overall sheath, to form a *cable*. A cable may have two or more *cores* – each core being a separately insulated conductor, except where the earth connection (protective conductor) is also carried within the sheath. A common form of cable consists of two cores and earth – a red-coloured insulated conductor for connection to the phase wire, and a black-coloured insulated conductor for connection to the neutral wire, together with an uninsulated (bare) protective conductor (*Figure* 2.1).

These three cores are usually laid in a flat formation surrounded by a sheath or outer casing, for protection.

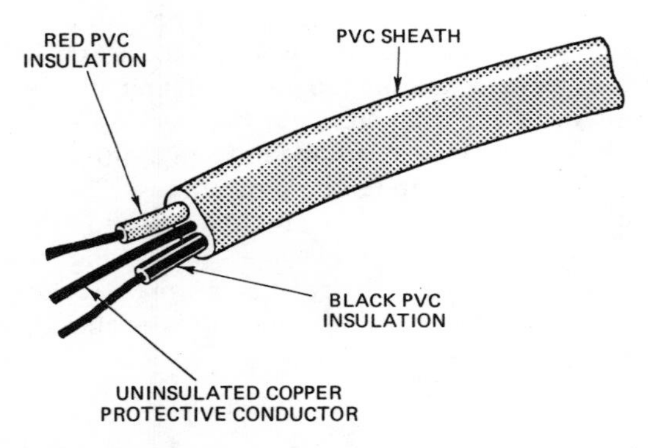

Figure 2.1. The protective conductor commonly incorporated in sheathed cable

The cable may be made up with a single core, which still has an outer sheath for protection against mechanical damage, or with any number of cores. In PVC-sheathed cables with three insulated cores, the cores are coloured red, yellow and blue.

Sheaths may be made up as follows:

Plastic sheathing

For PVC-insulated cores a PVC sheath is most often used. A variant is a polythene-insulated core or cores covered with a PVC sheath.

Tough rubber

Older cables are known as TRS (tough rubber sheathed). Rubber sheathing is usually employed with rubber-insulated cores.

Mineral-insulated (MI) cables
These cables, though more expensive than the flexible sheathed type mentioned above, are capable of being used in situations where no other cable could be employed.

Single-strand copper wires are embedded in a tightly compressed white powder insulation within a copper or aluminium sheath. The powder is made of magnesium oxide (see *Figure* 2.2).

The MI cable can be subjected to the full heat of a blowlamp without damage. With an overall covering of PVC the cable may even be buried direct in the ground.

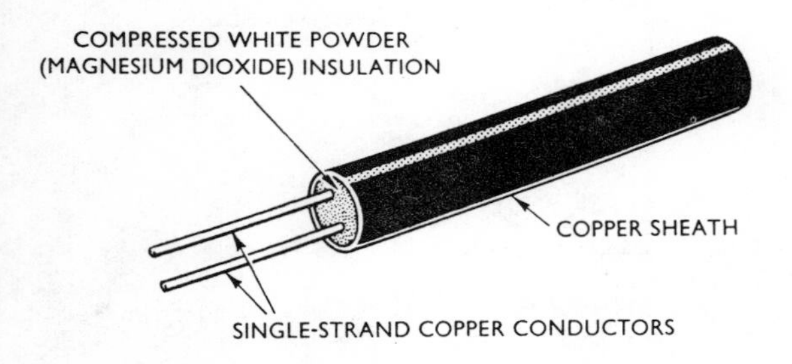

Figure 2.2. Mineral-insulated copper-sheathed cable

The only disadvantage (apart from cost) of the MI cable is that great care must be taken when it is cut and made off into a terminal. This is because the magnesium oxide is hygroscopic – in effect, it attracts moisture, which can reduce the insulation value. Therefore the ends of any MI cable must at all times be kept properly sealed, using the special sealing equipment supplied by the manufacturers. When the cable is cut and made off, the simple instructions supplied by the makers must always be carefully followed.

The enclosure

Electric wiring is best protected against mechanical damage by being inserted in a suitable enclosure.

Heavy gauge, steel screwed conduit
In this system, a complete enclosure made of screwed piping, and including all termination boxes and joint boxes, is provided. The steel conduit is cut to length, bent if necessary, and screwed so that it provides a perfect seal where it enters a termination box or where it is jointed (*Figure* 2.3).

The conduit system means the use of a considerable number of special tools such as dies, bending machines, pipe

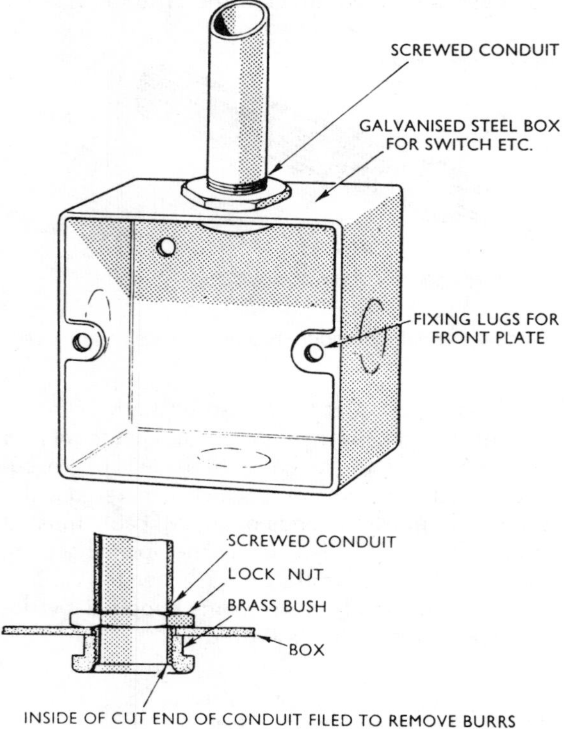

Figure 2.3. Typical box for the mounting of the switch or other fitment on a conduit system

vices, and saws. It is costly, and although it should always be designed in accordance with the Regulations so that the conduits are not tightly packed with wire, and draw-in points should be accessible, it is nevertheless not always easy to alter or extend a conduit system. A certain degree of skill and experience is needed before a conduit system is embarked upon, although there are no insuperable difficulties that could not be overcome by anyone used to handling metal-working tools.

Other types of conduit

Aluminium alloy conduits are sometimes used, and these must be screwed at joints and installed in the same way as steel conduits. They are subject to corrosion when embedded in cement and plaster, but may be protected by bitumastic paint.

Non-metallic conduit systems are also widely employed. These are made from a plastic tube which can be bent round corners. Some systems use screwed ends to fit into joints and terminal boxes, other systems employ cemented joints. This system is particularly suitable for installation where there is a strong possibility of corrosion, but where mechanical damage is not likely. Each circuit wired via this conduit must be connected to earth by a separate wire (protective conductor), with a covering of green/yellow PVC pulled in with the current carrying cables.

Protective channelling

The conduit systems mentioned earlier are used for the highest class of work but domestic installations usually consist of PVC-insulated PVC-sheathed cables which are normally concealed in the plaster. It is quite common, but not compulsory, for such cables to be enclosed in conduit or metal channelling. (See *Figure* 2.4.) This gives protection and also permits future withdrawal or renewal of the cables.

Steel and plastic trunking

Where a large number of cables have to be run in one direction, steel or plastic trunking, often of square section

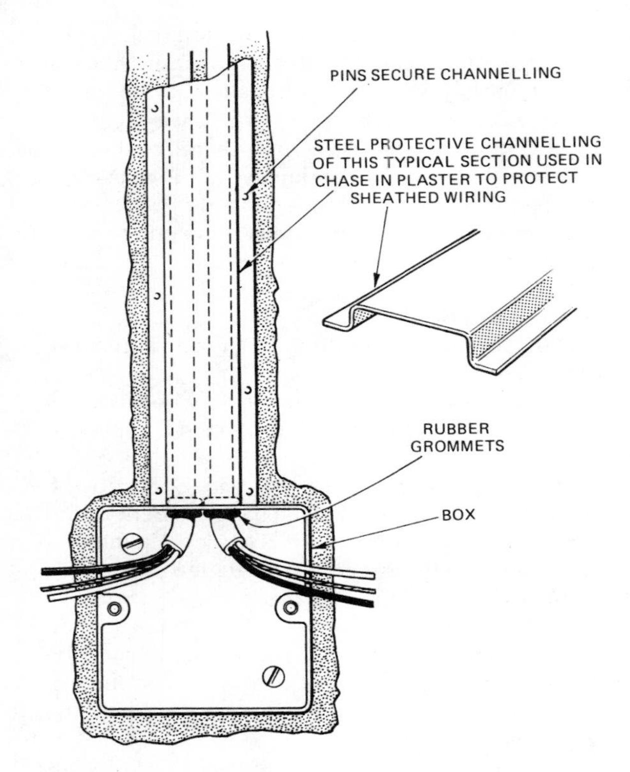

Figure 2.4. Installing sheathed cable in plaster, with protective metal channelling

with a screwed-down or clipped removable lid is used (*Figure* 2.5). Steel trunking must be electrically continuous and bonded securely to earth. Plastic trunking must have an insulated protective conductor (formerly known as earth wire) coloured green and yellow between all apparatus connected to the system.

Trunking is a very flexible system of installation. Suppose, for instance, a small business were started in a workshop, where initially only a few lights and one or two machines

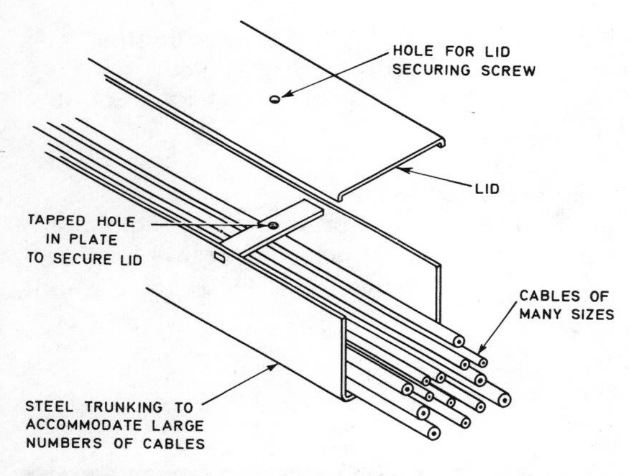

Figure 2.5. Where a number of cables follow the same route, steel or plastic trunking may be used to protect them, but the number of cables in each run is limited by the Regulations

were required. This installation could be completed using PVC cables in conduit, at moderate cost. If trunking were used, run at a high level on the walls around the room, the positions of lights, switches and machines could be fed using conduits, connected to the trunking and run down the walls to each position.

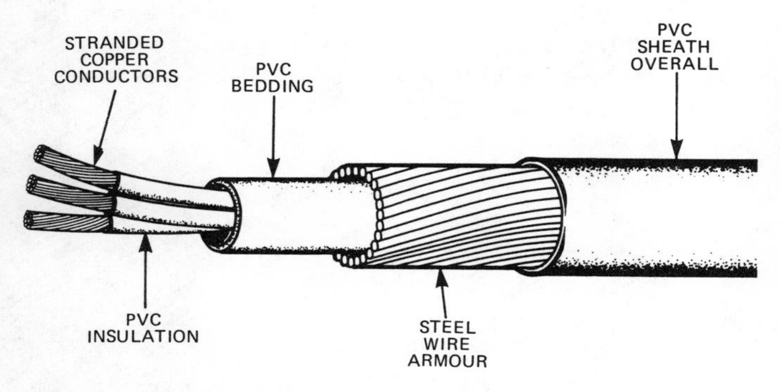

Figure 2.6. PVC steel wire armoured PVC cable

Though much of the installation might be superfluous at first, there would be plenty of room in the trunking for future additional wiring in the case of business expansion, requiring extra lights and machines, sockets, etc.

PVC steel wire armoured PVC

This cable has stranded copper (or sometimes aluminium) conductors with PVC insulation, enclosed in galvanised steel wire armouring and an outer covering of PVC (*Figure* 2.6). It

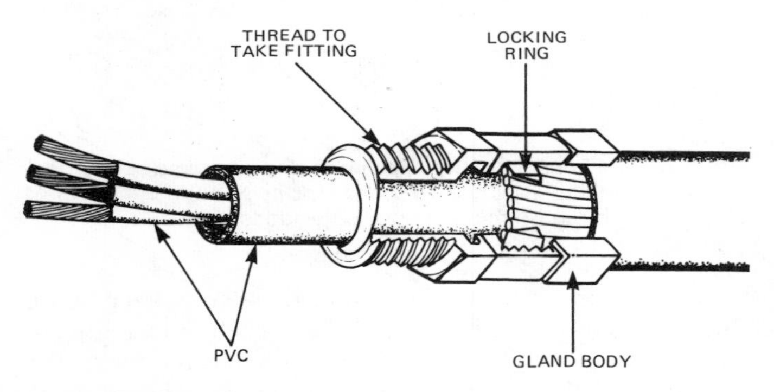

Figure 2.7. PVC/SWA/PVC gland

is often used for underground supplies to outbuildings, laid in fine earth or sand and covered with protective tiles. The cable may be fixed onto most surfaces using plastic clips; terminations require glands which are simple and quickly made off to the cable ends (*Figure 2.7*).

3

The electrical layout

As mentioned earlier, all electrical systems depend on there being a complete circuit from the sources of supply to the appliance, such as a lamp, that is to be fed with current.

In an installation supplied in the ordinary way from the electricity board's mains, the source of supply is the supply terminals on the consumer's fuseboard.

Every circuit must start from the phase fuse terminal and return to the neutral terminal.

Circuits are interrupted so that the power may be controlled by:

switches
fuses
miniature circuit breakers (MCBs)
contactors and relays
time switches
thermostats

Switches

The simplest circuit consists of a pair of wires from the mains terminals supplying one appliance, say a lamp.

In this circuit there must be a switch, and that switch must be situated in the phase wire, as shown in *Figure* 3.1.(a), not in the neutral wire.

This requirement, which is of course part of the Regulations, is especially important. If an appliance, for example, a

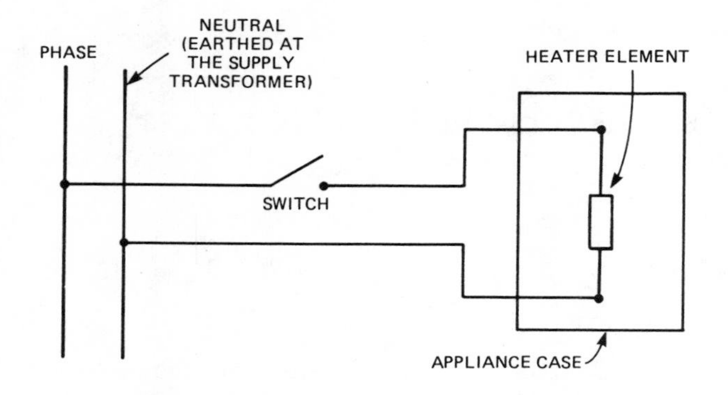

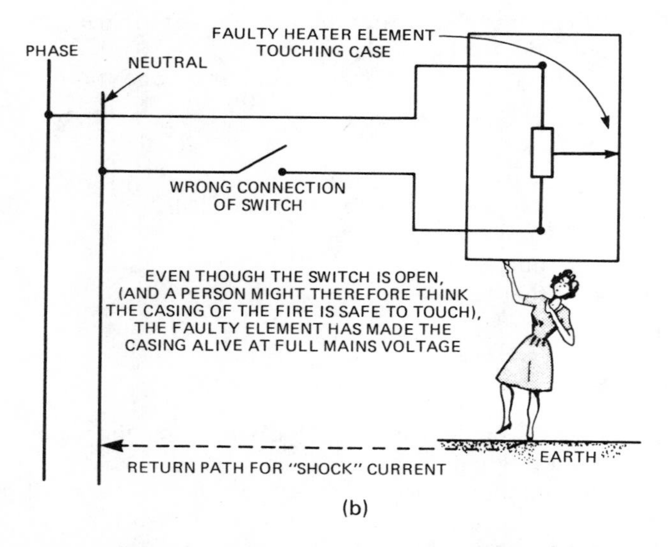

Figure 3.1. The switch must always be installed in the phase wire, and not in the neutral. The lower diagram shows what happens if the switch is wrongly installed and an appliance becomes faulty

vacuum cleaner, is connected to a switched socket-outlet, and the plug is left inserted, the appliance may not be in operation because the switch on the socket is 'off', and therefore any uninstructed person might start to tinker with it, thinking that because it is not running, it is dead. But if the switch were placed so that it interrupted the neutral (black) wire and not the phase (red) wire, the phase wire of the supply would be carried through the flexible cable to the appliance, even if it is not running, and a fatal shock might result from anyone interfering with the appliance.

The next variant of the simple one-lamp circuit is the two-way switch, shown in *Figure* 3.2, used for example on staircases so that a person can switch off the upstairs light after going downstairs.

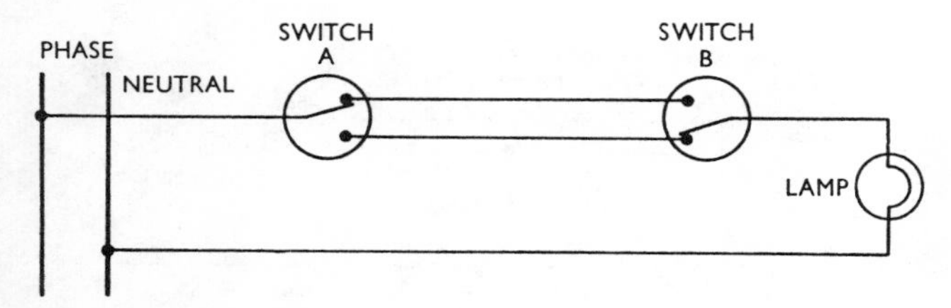

Figure 3.2. The two-way switch

In this circuit, the switches can have two positions, either of which can light the lamp. Suppose switch *A* is in the upper position, and switch *B* is in the lower position as in the figure. There is no circuit, so the lamp is out.

Now imagine a person near switch *B* turns that switch to the upper position. A circuit is established, and the lamp lights. He goes downstairs and reaches switch *A*, and wishes to turn the lamp out. He brings switch *A* to the lower position, and the lamp is extinguished. A second person, at switch *B*, has only to move his switch to the lower position for the lamp to light once more.

When two-way switching is being used care must be taken to see that the two-way type is bought. Some switches, e.g.

certain cord-operated types, are supplied by some manufacturers in the two-way version only, but a two-way can of course be used as single-way.

Suppose now that on a long staircase, for example, with several landings or a long corridor or a room with three entrances, it is desired to arrange for the light to be switched on and off at several points. In this case intermediate switches are used.

The two wires between the switches *A* and *B* in *Figure* 3.2 are called the strapping wires. If, in the case shown, the wires were to be reversed, a circuit would be established and the lamp would light.

The intermediate switch shown in *Figure* 3.3 carries out this reversal of the strapping wires, and any number of intermediate switches may be installed.

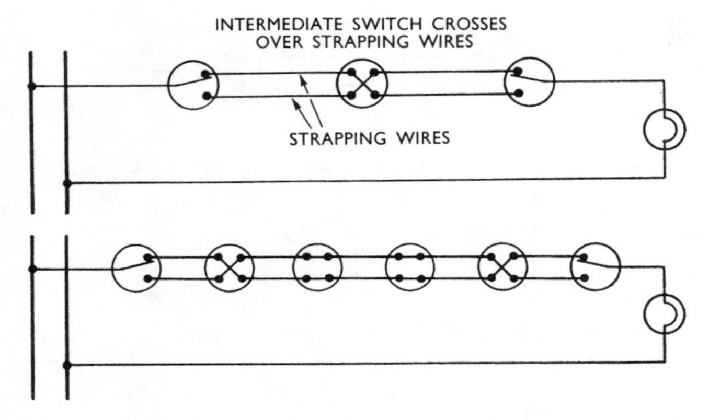

Figure 3.3. The use of an intermediate switch to control a lamp (or other load) from a number of positions

In *Figure* 3.3 the upper diagram shows a single intermediate switch but in the lower diagram there are four such switches. The two positions of the switch contacts can be seen from the two diagrams. In the circuit shown in the lower diagram, the lamp can be turned off and on from six positions.

Fuses

Fuses are found on domestic installations at three points.

It should be made clear that with normal, standard supplies only the phase wire is fused. The neutral connection simply has a link. It is totally incorrect, and in contravention of the Regulations, to fuse the neutral side.

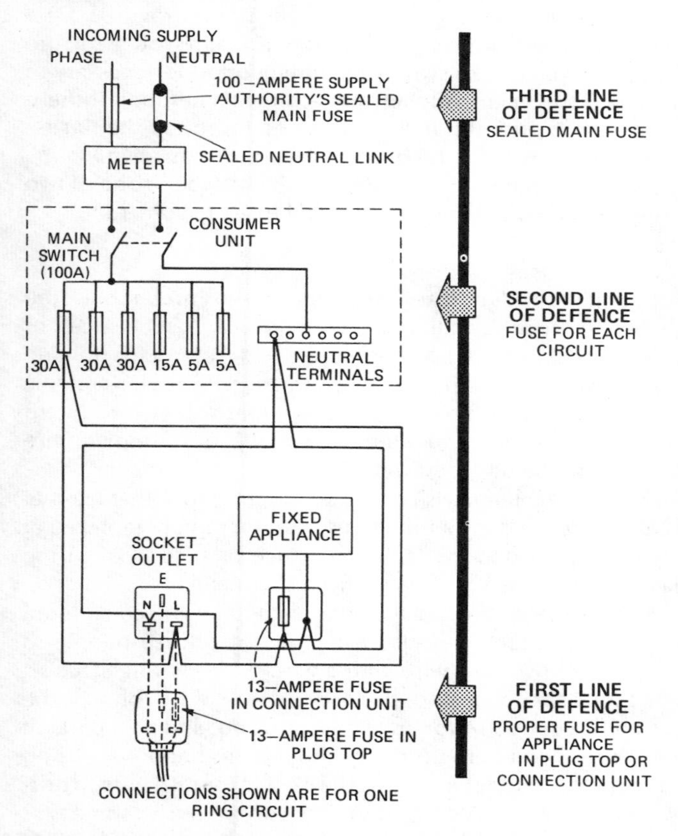

Figure 3.4. Fused protection on a domestic installation showing the three lines of defence against the consequences of faulty apparatus. (Protective (earth) conductor not shown)

There are the board's main fuses, or cut-outs as they are called (*Figure* 3.4). They are the third line of defence. These are sealed, and must not be interfered with. If the main fuse blows, the electricity board's engineers must be called out to replace it.

Secondly, there is the consumer's unit, with a fuse or MCB for each circuit or group of circuits, and a main switch for that fuseboard. (*Note:* this is the only switch provided. The electricity board no longer provides a main switch for the whole installation. To switch everything off, it is necessary to switch off the main switches on each consumer unit, if there is more than one. But it should not be forgotten that inside each, the wires from the board's meter are still alive.)

The fuses used in the consumer's unit may be of one of two types: rewireable fuses, or cartridge fuses (*Figure* 1.6).

Rewireable fuses

Although cartridge fuses are considered preferable, the rewireable type is still quite common. A fuse bridge, of non-flammable material, is equipped with screws for holding the fuse wire. This wire is usually threaded through a ceramic tube, or otherwise held in some way that will ensure that if the wire gets hot and ultimately melts or fuses, no fire damage can result (see *Figure* 1.6).

The rewireable fuse is convenient in the sense that if a reel of fuse wire is handy a blown fuse can be rapidly replaced at any time, provided the faulty appliance or section of wiring has been repaired or isolated from the mains.

Perhaps the rewireable fuse may sometimes be considered as too convenient, because it is easy for unwise people to replace a blown fuse wire with a wire of the wrong size.

If a 30A circuit is properly fused, the fuse will blow if the current exceeds 30A for any length of time, and if all parts of the circuit wiring are correctly chosen, no harm will result. But if the fuseholder is renewed with thicker fuse wire, which will allow, say a 50A current to flow continuously, some part of the circuit – perhaps a switch, or a flexible cable, or a socket-outlet – may become overheated and a fire could possibly result.

When replacing a fuse wire, always make sure that the screws are properly tight, but not too tight to damage the wire and reduce its current-carrying capacity. Check that the wire is properly fitted into the safety tube or path through the fuseholder. Do not strain the wire too tightly between the terminals, as tightening up may stretch it and reduce the copper section.

Fuseholders are usually marked with the *maximum* size of fuse wire they should carry. This figure should never be exceeded.

Cartridge fuses

The best type of fuse is one in which the actual fusible element is enclosed in a flame-proof cartridge. In some cases the cartridge is filled with a type of sand intended to extinguish any flame that might result from a fuse blowing as a result of a heavy excess of current.

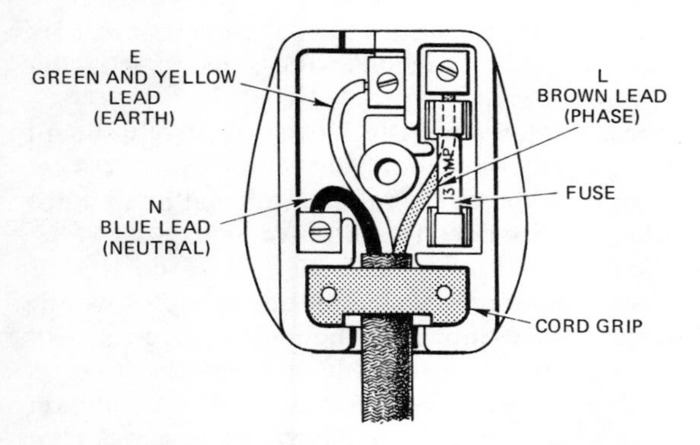

Figure 3.5. The fused plug top and the method of connection

The cartridge fuse is made in various sizes, such as 5, 10, 15, 20, 30, 60 A and so on.

It is slightly more expensive than the rewireable type, and there is always the problem of being caught with no spare cartridge of the correct size. Do not try to repair a blown cartridge fuse.

The best consumer's intake point installation will consist of cartridge equipment, each circuit being properly labelled as to its destination and the size of fuse needed, with an adequate supply of cartridge fuses in a convenient spot nearby.

The first line of defence in the household wiring system is the fuse to be found in the plug top in the 13 A plug shown in *Figure* 3.5.

This plug, now the recognised standard (British Standard 1363) has three 'square' pins (rectangular, to be precise) and the phase side is connected through a fuse. The fuses commonly available are: 3 A, red; 13 A, brown.

Care should always be taken to ensure that the correct size of fuse is inserted in the plug to suit the appliance to which it is connected.

Miniature circuit-breakers

As an alternative to fuses, small automatic circuit-breakers are now available, which are much the same size as the equivalent fuse (*Figure* 3.6).

A circuit-breaker is an automatic switch. It is so arranged that if a current of greater value than that for which it is set should pass through the device, the switch will open automatically and cut off the circuit, so preventing damage in the same way as a fuse.

The automatic operation of a circuit-breaker is known as 'tripping' – the switch trips a catch that holds it in. Trip mechanisms take two forms, both of which may be used on the same switch. An electromagnet – a coil of wire wound on an iron frame – has a mechanical pull that corresponds to the current passing through its coil. The current through the switch is taken through such a coil, arranged with an arm, attracted by the magnet, that can flick the switch to the off position if the current is too great.

In the second type of trip mechanism, a bimetal strip is used. All metals expand when heated, some more than others. If two strips of metal, one with a high rate of expansion with heat, and the other with a low rate, are joined only at their ends, when heat is applied the combined strip will bow, or bend, as the different rates of expansion come into play. Inside the switch there is a very small heater element that carries the main current. If the current is excessive, this element gets hot and causes the nearby bimetal strip to bend, and a linkage then trips the switch.

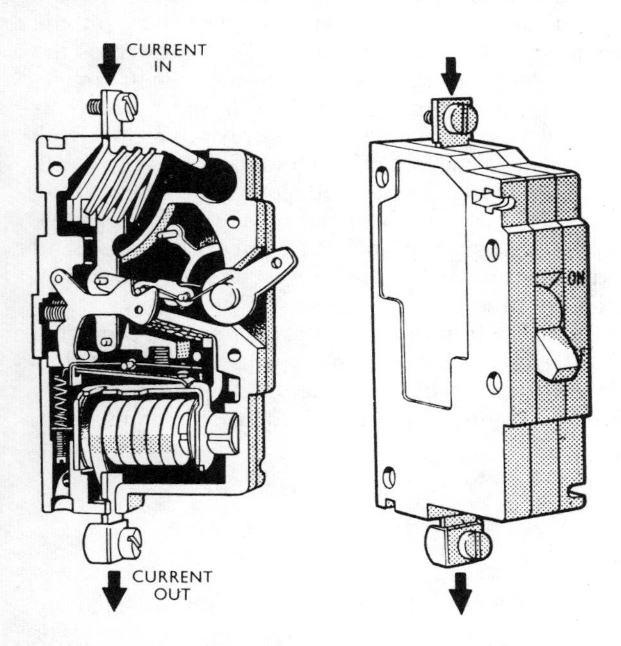

Figure 3.6. The miniature circuit-breaker, often used in preference to a fuse (reproduction of a Crabtree design)

With the electromagnetic type of trip, the switch can be closed again immediately it has tripped. (It may trip again if the cause is still there.) But with the thermal, or bi-metal strip type, the switch cannot be closed again for a minute or two, as the heater element and the bi-metal strip have first to cool down.

Time switches

A time switch is often used for controlling such circuits as shop-window lighting and central heating. Like all other switches, the actual contacts of the time switch must be wired into the phase side of the circuit, and not into the neutral, and care must be taken that the rating of the time-switch contacts – for example, 10 A, as indicated on the nameplate – is not exceeded by the appliances on its circuit.

Time switches may be of several kinds. In some, known as the spring-rewind type, an electric motor winds a clockwork system, so that if the supply should fail, the clock will continue to run, and will open and close the contacts at the proper time, for a period which may be a few hours or a few days. In other designs, the clock is electrically operated, and will stop if the mains supply fails; it must be set to the correct time when the supply is resumed.

In both cases, there must be a separate, fuse-protected circuit to the clock motor. Some clocks have the motor-circuit fuse incorporated in the case. In other instances, it is necessary to provide a 2 A fuse and a separate connection.

Thermostats

Thermostats are temperature-operated switches, which are arranged to open or close a circuit as the temperature rises or falls. For room heating control, they may work in the range of 10° C to 20° C, while for refrigerator applications they may operate in the range of 0° C to −20° C.

In many cases, bimetal strips, mentioned earlier, are used. As the temperature rises, the bimetal strip heats and bends (*Figure* 3.7) so that ultimately the contact is broken and the heater circuit switched off. To prevent the contacts 'dithering' and consequently arcing, the contact piece is usually equipped with a small magnet, and as the bi-metal strip slowly bends it is suddenly snapped to the open or

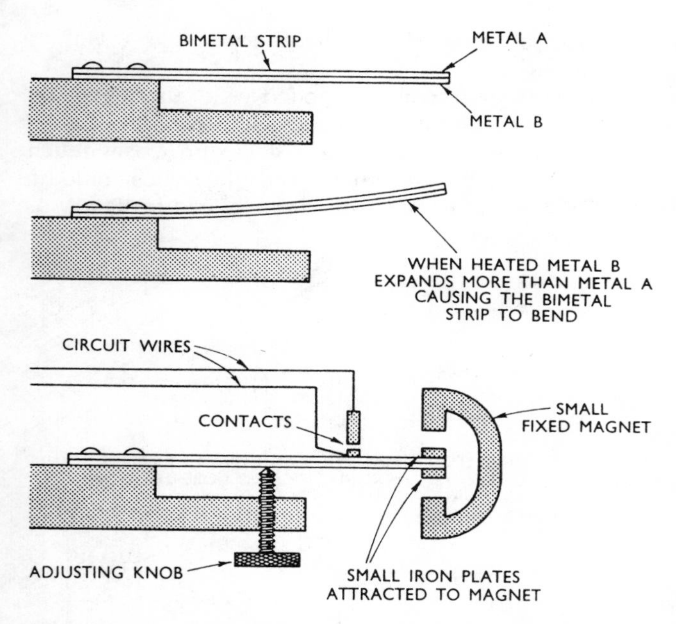

Figure 3.7. A refrigerator thermostat – the unit switches on as the temperature increases

closed position, as the case may be, thus giving a clean break.

Thermostats should always be wired into the phase side of the circuit, and again the current rating of the contacts should never be exceeded.

Relays and contactors

Devices such as time switches and thermostats have switching contacts that are limited in their capacity, usually to circuits of about 3 kW. If the time switch, for example, is to be used for larger circuits, some auxiliary device is necessary.

Small devices of this kind are called relays, larger examples are known as contactors.

An electromagnetic coil is supplied with current by the closing of the time-switch contact, and when this coil is energised it attracts an armature that in turn closes much larger contacts than those with which a time switch could be fitted: in fact, a contactor could control a load of 100 kW or more (see *Figure* 3.8).

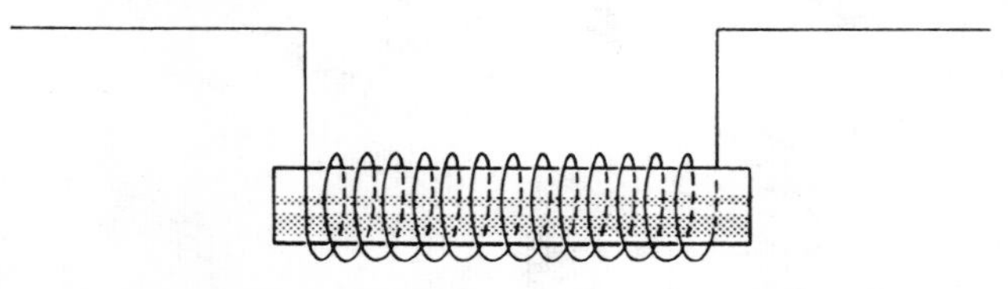

PRINCIPLE OF THE ELECTRO MAGNET. WHEN CURRENT PASSES THROUGH THE COIL WOUND ROUND THE IRON CORE THE IRON BECOMES MAGNETISED

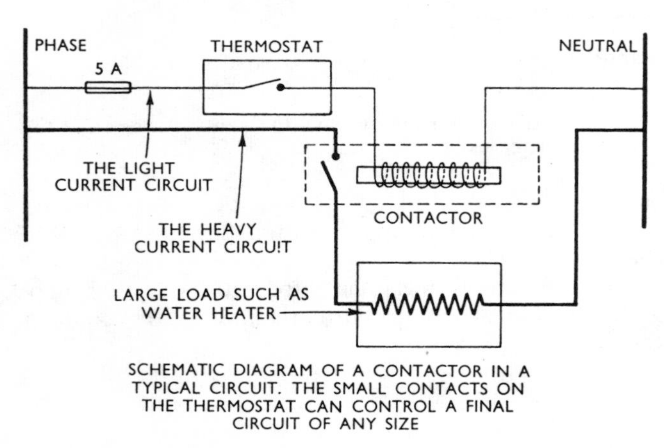

SCHEMATIC DIAGRAM OF A CONTACTOR IN A TYPICAL CIRCUIT. THE SMALL CONTACTS ON THE THERMOSTAT CAN CONTROL A FINAL CIRCUIT OF ANY SIZE

Figure 3.8. The principle of the contactor

A relay is a smaller version of the contactor, and is usually of the type where the originating current is very small, such as the output from photoelectric cells which automatically switch on lighting equipment at dusk.

In connecting up contactors and relays, care must be taken to fuse the control circuit with a light fuse (a 5A fuse is usually sufficient) and then to see that the fusing on the main circuit is adequate.

Both the control circuit and the main circuit switches must be wired into the phase side, as for all switches or circuit-breaking devices.

The circuit layout

We can now consider the circuit layout for a typical installation, for example, a four-bedroomed house.

Bearing in mind the first principle that each circuit must have proper protection by being adequately fused, one's mind must first turn to a system whereby every light, every appliance (such as a fixed radiator or cooker or refrigerator) and every socket-outlet had its own separate cable back to the main consumer's fuseboard, and was individually connected to a separate fuse of the appropriate size.

This would indeed be the perfect and ideal system, but would be extremely expensive on account of the lengths of cable needed and the number of fuses required. It is also unnecessary.

To consider the extreme alternative, suppose an installation had only one main fuse, to which all the circuits were connected. If this fuse blew, through a fault of any one appliance, such as a desk lamp, the whole house would be plunged in darkness. In any case, proper *graded* protection for the various circuits could not be provided in this way.

The practical solution obviously lies somewhere between these two extremes.

The number of circuits that may be grouped together on one fuse depends on the total demand on the circuit, evolved according to certain rules. For example, consider a lighting circuit. This might well originate at a 5A fuse in the consumer's unit. It is laid down that each fixed lampholder must be assumed to carry a 100W lamp. Now a 100W lamp consumes 0.416A. So 12 lampholders could legitimately be supplied from a single 5A lighting circuit – *providing the correct size of cabling was used throughout the circuit.*

But this would not be practicable in the house we are discussing, since the wiring would become somewhat cumbersome and in any case it would be undesirable for every light in the house to go out if a single lampholder developed a fault.

A commonly used system would be to have two 5A lighting circuits, one for the first floor and one for the ground floor, with perhaps ten or so lampholders wired to each. In this way there would be some light left in the house if one lampholder failed, and in addition the circuits would be a little underloaded so that extensions would always be possible.

Turning now to fixed appliances – cookers, water heaters, fixed radiators, and refrigerators – some of these (except perhaps refrigerators) are heavily loaded appliances, often calling for 12kW or more in the case of cookers, 3kW for water heaters, and so on.

There should be a separate final circuit for every appliance rated at 15A and above, except in the special circumstances mentioned later when the ring circuit is discussed.

Therefore each of these larger fixed appliances should in fact have a separate circuit back to its own fuse, of appropriate size, in the consumer unit. Even if some appliances, such as a particular fixed radiator, do not initially require more than 2kW (8½A) it is strongly recommended that the circuit should be wired back to the consumer unit on the basis of a single feed of 15A capacity: no one can tell if at some time in the future a larger radiator may not be needed at that point. Experience shows that the usage of electricity steadily grows, not only for installing *more* appliances but in substituting new, more heavily loaded appliances for older ones. For example – to mention a portable appliance – the elements fitted to electric kettles have grown in loading from the initial 600W, and have now reached 3000W or even more.

The ring circuit

Anyone who thinks carefully about the circuit arrangements set out above will soon reach the conclusion that there is some inevitable wastage of copper – that is, there is more current-carrying capacity than is needed *for most of the time*.

On the whole, it is better to have a little extra capacity in the circuits, to allow for extension. But there is a very widely used system that allows much fuller use to be made of the current-carrying capacity of the circuits installed.

This system is called the ring circuit (*Figure* 3.9), and it is based on the application of what is called the principle of 'diversity'.

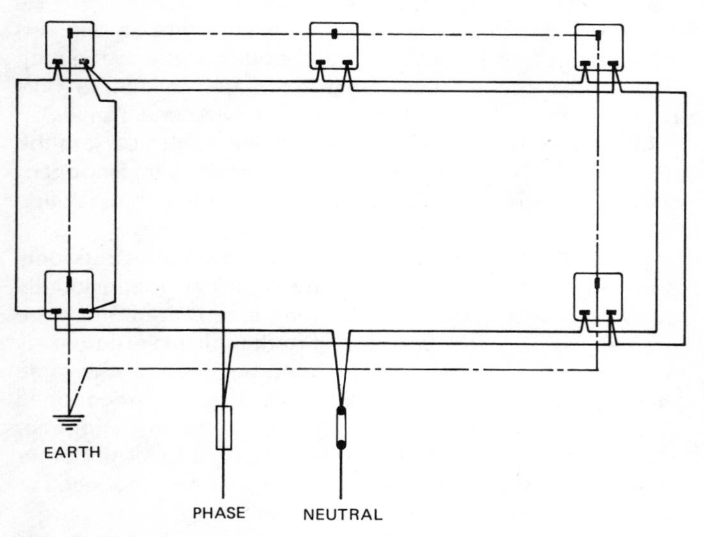

Figure 3.9. The principle of the ring circuit

Take any ordinary house, and consider the time and place at which various appliances are used.

The largest portable electricity-consuming appliance commonly used in domestic premises is the 3 kW fire. It is extremely unlikely that more than three such fires will be in use, at full load, at the same time, even in the coldest weather, at any rate on the ground floor.

Therefore, if there are, say, eight socket-outlets on the ground floor, each of 13 A capacity (thus capable of taking the

3 kW fire) it would be extremely generous in copper to wire each one separately back to the consumer unit. When the three large fires are in use, the remaining five sockets on the ground floor are likely to be used only for very light current appliances like lamps, radio and television sets.

Now suppose all the 13 A standard socket-outlets on the ground floor were connected in a ring. That is, a pair of wires starts at one 30 A fuse at the consumer unit and runs to the first socket, on to the second, the third, and so on, and then back to the *same* 30 A fuse (see *Figure* 3.10).

The ring circuit is based on the employment of a standard socket-outlet, of 13 A capacity, and having a fuse in the plug top. This socket-outlet will allow for an appliance up to 3 kW being connected, as such an appliance will need 12.48 A.

In this system, each 13 A socket has two routes back to the mains. The *maximum* use is made of the *minimum* amount of copper in the cables: that is, the minimum length of cabling is employed for a given number of socket-outlets.

The use of the ring circuit, as mentioned above, is only possible because of the diversity of usage that naturally evolves. In an ordinary dining room, where there might be perhaps six 13 A standard socket-outlets, it is very unlikely that more than two 3 kW fires would be in use at the same time, even under arctic conditions. Fires, as mentioned earlier, are the largest portable current-consuming appliances likely to be used in domestic premises. Any other appliances use so much less current that they do not need to be taken into account here.

The Regulations regarding the use of the ring circuit are:

If the floor area concerned does not exceed 100 square metres there can be an unlimited number of 13 A standard socket-outlets on the ring, which must consist of 2.5 mm^2 (minimum) conductors (assuming PVC-insulated cable) and must terminate in a 30 A fuse. The cable size may be affected by the protective conductor requirements.

Note: In practice, in domestic premises, the requirement mentioned above may mean that two separate rings are needed. A convenient division is to have a downstairs ring

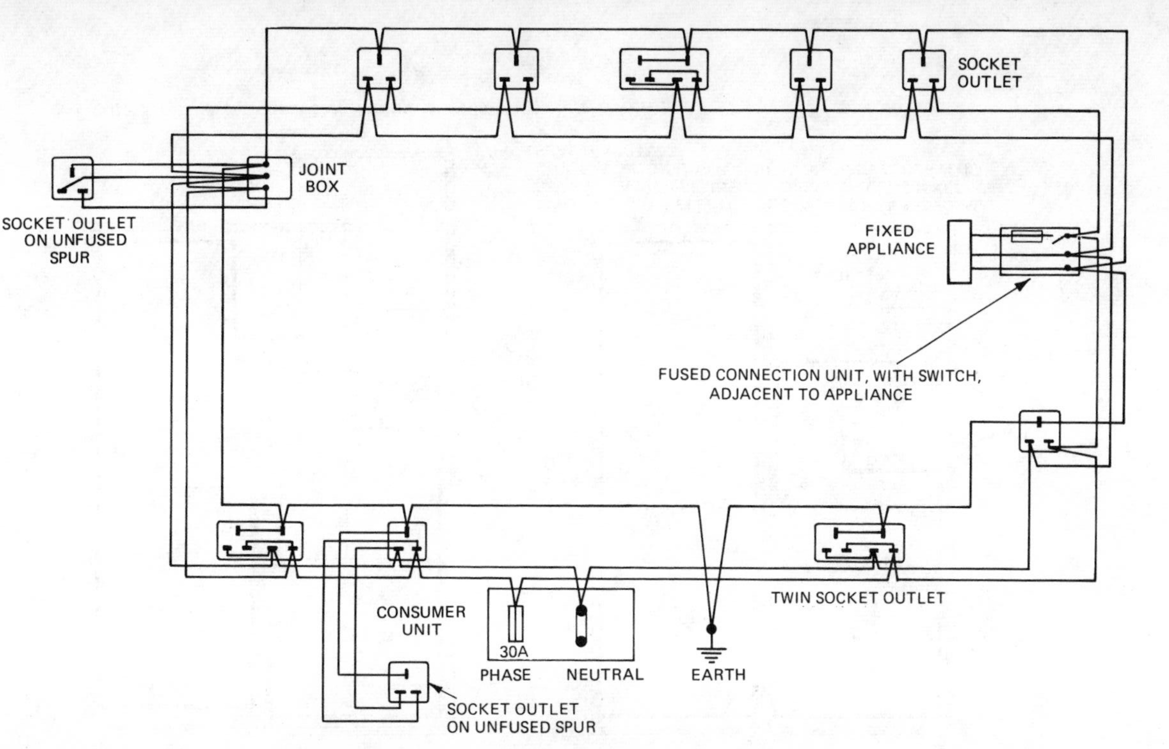

Figure 3.10. The ring circuit in greater detail

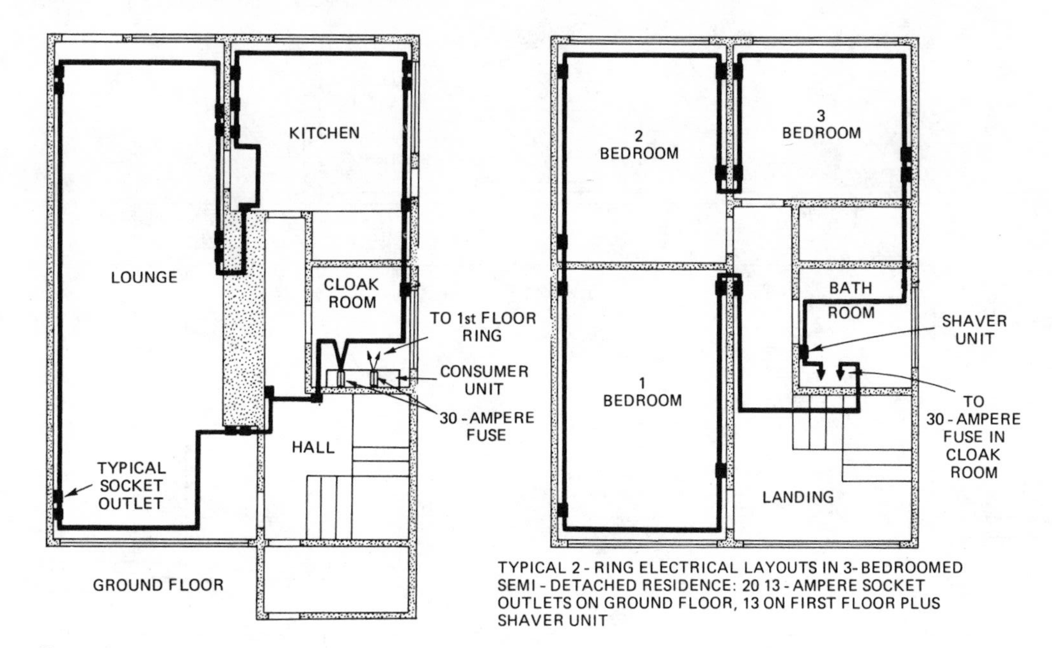

Figure 3.11. A practical ring circuit installation, using two rings

and an upstairs ring (see *Figure* 3.11). The more the rooms embraced by the ring, the greater the diversity. (People do not, in general, use all the rooms in a house at the same time, and even if they did, their current-using habits do not coincide.)

In addition to 13 A socket-outlets, it is permissible to install specially designed connections for small-current appliances such as shavers. Shaver outlets must comply with BS 3052 in bathrooms or BS 4573 in other locations.

Appliances consuming above 3 kW, such as cookers and large immersion heaters, should be wired back separately to appropriate fuses in the consumer unit. This applies to immersion heaters fitted to storage vessels of more than 15 litres capacity. See IEE Appendix 5.

It should be noted that special regulations, to be mentioned later, apply to the method of connection of fixed appliances to the ring.

Spur connections

There are two types of spur connection that may be made to a ring circuit – unfused spurs and fused spurs. This avoids the need to run the four wires of the ring to an isolated area, providing the following points are observed:

When unfused spurs are connected to a ring circuit (*Figure* 3.12), not more than one single socket, or one twin socket, or one fixed appliance shall be fed from each, and the total number of unfused spurs shall not exceed the total number of socket-outlets and stationary appliances connected directly to the ring. Unfused spurs shall be connected to a ring circuit at socket-outlets, or in suitable joint boxes.

The conductors supplying the unfused spur shall not be smaller than those forming the ring itself.

The fused spur (*Figure* 3.13) is connected to the ring via a fused connection unit, the rating of the fuse in the unit not exceeding that of the cable forming the spur, and in any event not exceeding 13A. The connection unit is provided

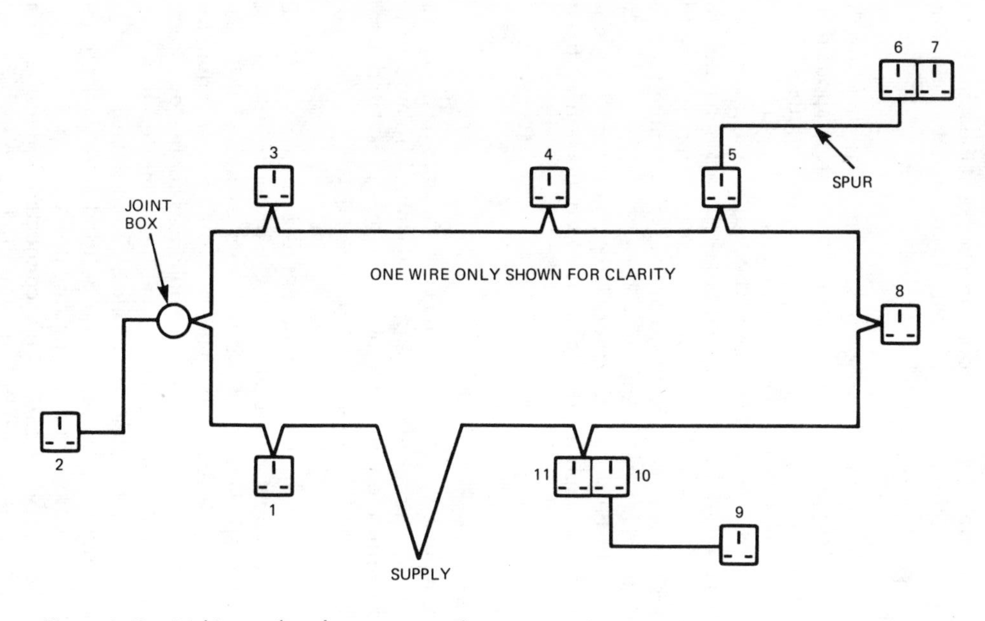

Figure 3.12. Making unfused spur connections

with a cartridge fuse which can be replaced easily, usually from the front of the box. It is useful in cases where a fixed appliance, such as a wall-mounted fire, needs to be supplied. The use of this unit obviates the need for wiring back to the consumer unit. The conductors on the fused spur need only be of a size suitable for the fuse fitted within the connection

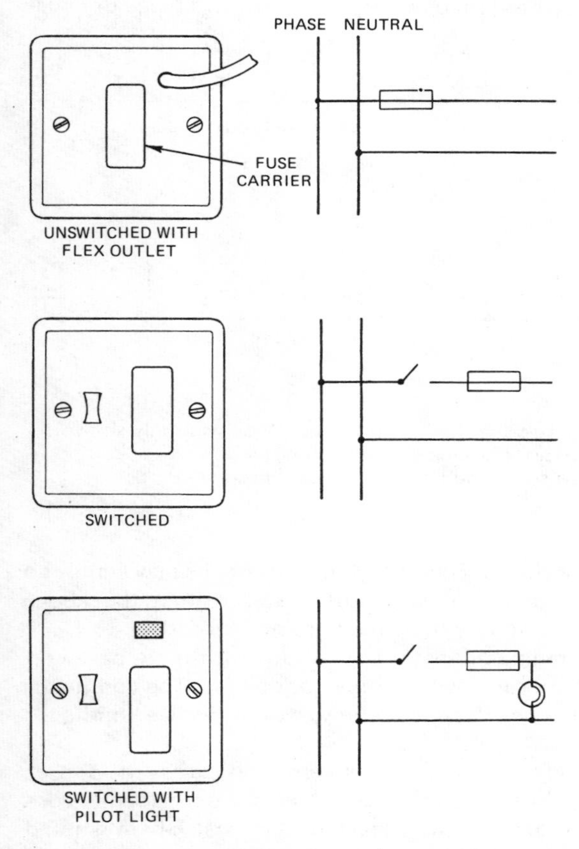

Figure 3.13. Typical fused connection units, showing the connections

unit, except that if the spur supplies a socket the cable must be a minimum of 1.5 mm^2 (assuming PVC-insulated cable).

Radial circuits

Radial circuits are circuits which also utilise 13 A flat pin plugs and sockets, to British Standard 1363, except that the circuit is not wired in the form of a ring. There are two types of radial

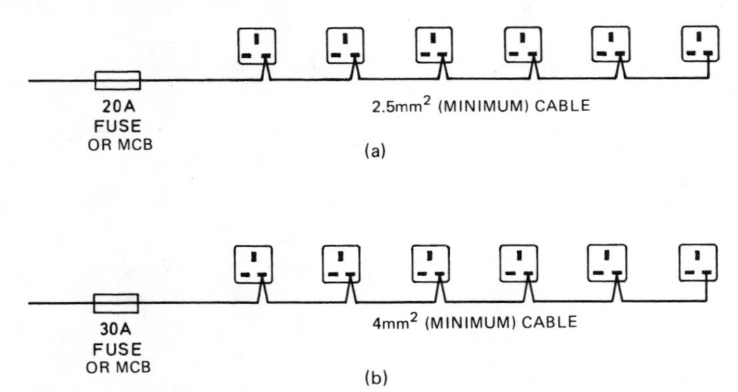

Figure 3.14. Two types of radial circuit (phase wire only shown)
(a) Unlimited socket-outlets serving 20 m^2 maximum
(b) Unlimited socket-outlets serving 50 m^2 maximum

circuit, as shown in *Figure* 3.14. Radial circuits may supply an unlimited number of socket-outlets, provided the above floor areas served by the circuit are not exceeded.

The minimum conductor sizes shown above are based on PVC-insulated cable with copper conductors. The conductor size may also be affected by the size of protective conductor required.

Immersion heaters supplying vessels in excess of 15 litres capacity, or heating appliances which are permanently connected and form part of a space heating installation, should NOT be connected to ring or radial circuits using socket outlets to BS 1363. IEE Appendix 5 refers.

Off-peak circuits
In this system, the area electricity board provides a time switch to allow the off-peak supply to be taken at the specified hours only. Sometimes there may be provided, in addition, a contactor to switch on the off-peak circuits, since the time-switch contacts may not be of large enough current-handling capacity. The consumer himself has to supply the extra consumer unit, if required, allowing for the required number of fuses (often termed 'ways'), and the requisite off-peak circuits.

Off-peak supplies are generally used in three ways: for thermal storage heaters, under-floor warming and storage water heating.

In addition, battery chargers are sometimes fed on the off-peak system.

The same appliances or equipment may be used on the normal or off-peak supplies and *all* consumption during the off-peak period is at the cheaper (off-peak) rate.

It is usual for appliances, such as storage heaters, which will be used on the off-peak supply *only*, to be wired to a separate consumer unit, connected to the off-peak supply. Items which may be used during either the normal or off-peak periods are wired to a consumer unit connected to the normal supply. Storage heaters are usually wired on individual circuits back to the consumer unit and controlled by a switch, with flex outlet, fitted adjacent to the heater. *Figure* 3.15 refers.

When the off-peak system is used, at the beginning of the off-peak period a sealed time switch energises a relay in the meter which transfers the meter drive from one set of recording numbers to another. At the end of the off-peak period the time switch de-energises the relay, switching meter numbers back to the daytime recording.

The off-peak period is normally 7 hours, during the night.

As stated earlier, it is usual for storage heating appliances to be wired to their own consumer unit or fuseboard. However, the consumer may use an additional time switch of his own, if desired, this time switch being the responsibility of the consumer, as it is quite permissible for any appliances

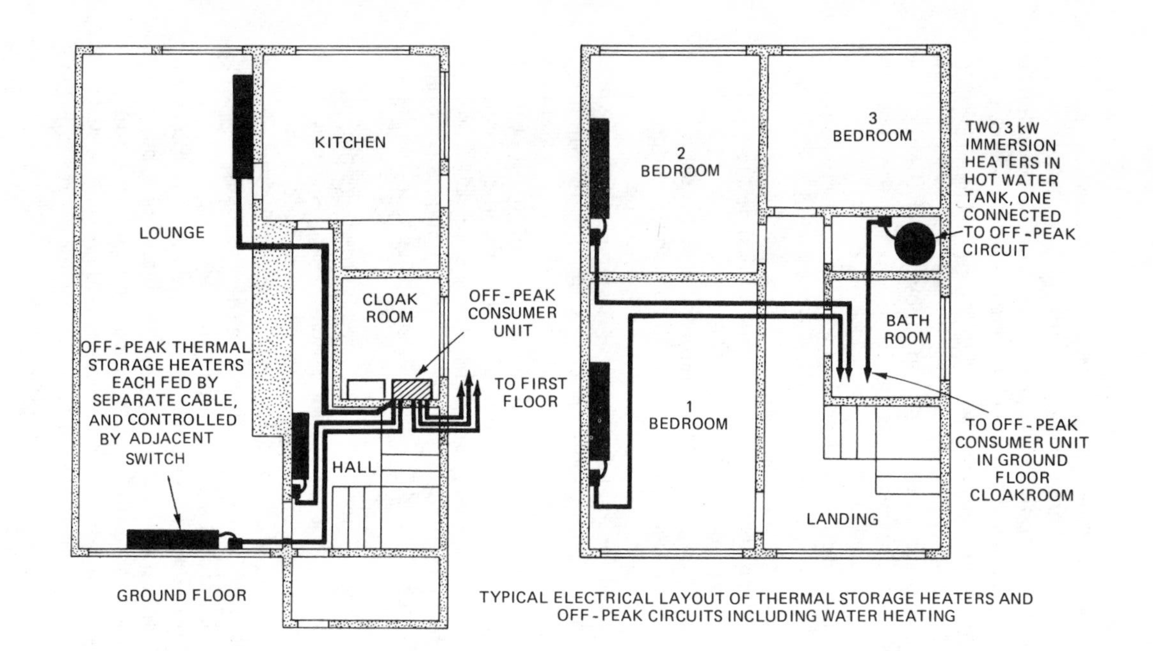

Figure 3.15. Layout of off-peak circuits for storage heating and water heating supplies in domestic premises

to be used at any time of the day, although a higher rate will be charged during the 'normal' period. For the above reason, it is in the consumer's interest to use electricity during the night and it may be cheaper to use, say an automatic washing machine connected by a time switch, in this period.

Summary

To sum up, the circuit layout in a typical house comes under the following headings:

Fixed appliances

(a) Cookers and immersion heaters, and any other fixed appliances using 3 kW or above (large current-consuming appliances).
Each appliance usually wired directly and individually back to the consumer unit.

(b) Other fixed appliances may be connected to ring or radial circuits via fused connection units.

(c) Lighting circuits for fixed lamps (light current-consuming appliances).
Grouped in circuits, each circuit wired back to the consumer unit.

(d) Off-peak circuits, for storage heaters, floor heating, immersion heaters.
Each appliance wired singly or in groups back to the consumer unit on separate circuits connected on off-peak supply.

Socket-outlets for portable appliances

The ring or radial circuit is used, with spurs as required.

Earthing

As mentioned earlier, earthing is a means of ensuring electrical safety. All metal parts of all appliances used in the installation should be connected solidly to earth at all times that the appliance is connected to live electric mains.

For fixed appliances, earthing is ensured by a permanent earth connection, which must be solidly connected to an earth point, a feature to be mentioned later. The connection must be made by means of (*a*) the steel conduit, if such a system is properly installed, or (*b*) a protective conductor of suitable size.

For portable appliances, fed by means of socket-outlets, the earth connection is ensured by means of the earth pin in the plug. The earth socket, into which this pin enters, must be connected to the earth point by proper permanent means, as in the case of fixed appliances.

In the majority of cases in domestic premises the earth socket is connected to the earth point by means of the uninsulated protective conductor in the sheathed cable itself.

The earth connection

This raises the question as to what is an 'earth' connection. A wire buried in the earth itself forms a kind of earth connection, but an earth electrode of this kind has to be very carefully designed so that it will not corrode or, for example, find the earth around it becoming so dry that the electrode becomes insulating instead of conducting.

Wherever practicable, the consumer's earth terminal is provided by the electricity board, although this is not obligatory. In the case of an underground service the incoming supply cable may have a lead sheath and/or steel armouring. This is connected to the earth electrode at the substation and to the consumer's main earthing terminal adjacent to the meter. All the earthing points of the installation should be connected back to this earthing terminal, which is the official earth point (*Figure* 3.16).

But if such a terminal does not exist, what can be done? In any case, overhead supplies, such as those provided in many rural areas, do not allow for the provision of an 'official' earth point.

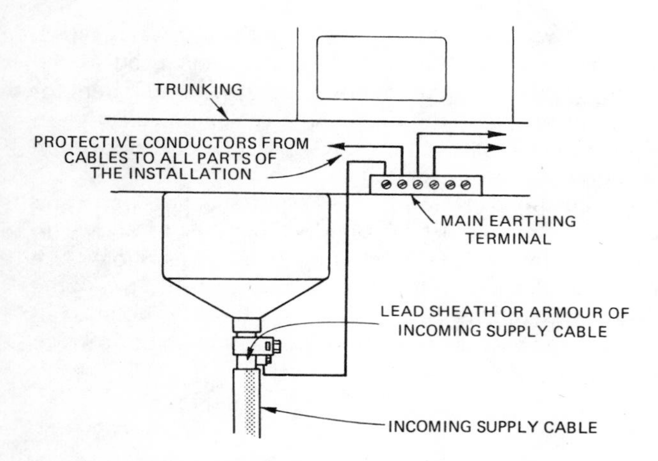

Figure 3.16. Earthing terminal at the consumer's intake point. *Note:* this terminal is not always provided by the electricity board, particularly in rural areas where the supply enters the house from an overhead power line

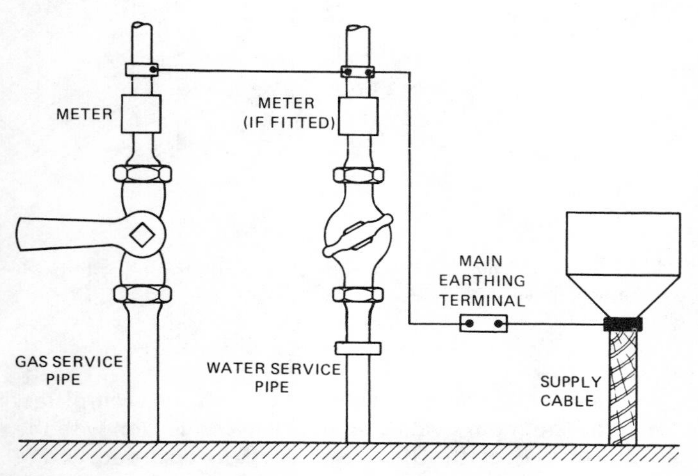

Figure 3.17. Bonding of services. Connection to gas service pipe should be within 600 mm of gas meter

The cold water system (*never* the hot water water pipes, these may be discontinuous) has often been used as the protective conductor system, however, this is insufficient for the sole means of earthing of an installation as modern systems often use plastic pipes. A separate earth connection *must* therefore be provided.

[The regulations (413-2 and 547-2) state that gas, water and electricity services must be bonded together as shown in *Figure* 3.17. The purpose of bonding the services is to ensure that all metal work within the premises is at the same potential, so reducing the risk of electrical shock, but it is not permissible to use gas or water pipes as the sole means of earthing.]

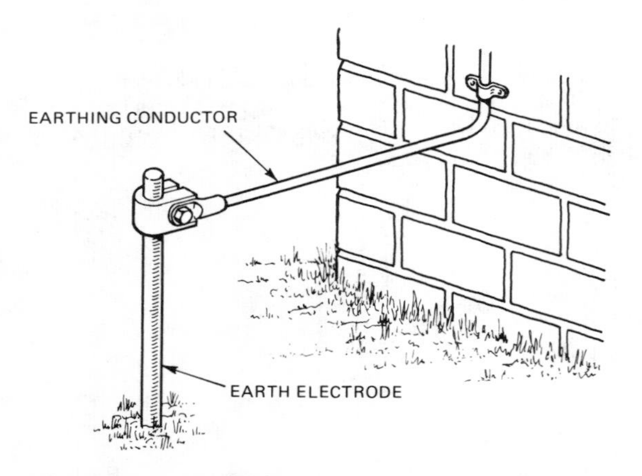

Figure 3.18. A typical earth electrode providing an earth point where no other is available

An earth electrode is a metal rod or rods, or even plates, set in the ground providing an effective connection with the general mass of earth. The size of the earth electrode, and of the earthing conductor which makes the connection from the electrode to the earthing terminal, should be of the correct size, calculated according to IEE Regulation 542-16 (*Figures* 3.18 and 3.19).

But even with this arrangement, we are not out of all our earthing difficulties where no 'official' earthing point is provided. The ground near the installation may be rocky, or exceptionally dry, or there may be some other condition that leads to a high resistance between the earth electrode and the general mass of earth.

This could give rise to dangerous conditions. The whole ground might become alive, in the neighbourhood of the earth electrode, at the time of a fault, and a child, for example, touching perhaps a rainwater down-pipe with a water flow into a gutter, could be killed by electrocution, the current passing *from the ground* through his body, to the rainwater pipe (*Figure* 3.20).

Earth leakage circuit-breakers

The problem of providing proper earthing facilities under these conditions is solved by the use of the earth leakage circuit-breaker (ELCB).

Suppose an earth spike, or electrode, is provided. Assume also that very dry or rocky ground means that its resistance to the general body of earth is rather high – say up to 100 Ω, instead of much less than 1 Ω, as it should be.

Now imagine a wire connected from the earth terminal of the wiring installation as a whole to this electrode. If then an appliance fails – say an electric iron becomes faulty internally – the phase wire will become connected to the earth electrode, through the green and yellow protective conductor on the iron.

Under ideal conditions this occurrence would immediately cause enough current to flow to blow the fuse, and any danger to the person using the iron would be averted.

But with our high-resistance earth electrode, enough current does *not* flow to blow the fuse. The body of the iron remains alive.

However, *some* current does flow, and it is this current that is used to safeguard the installation.

SAFETY ELECTRICAL CONNECTION

DO NOT REMOVE

Figure 3.19. Label to be fitted to earthing lead. Lettering not less than 4.75 mm high

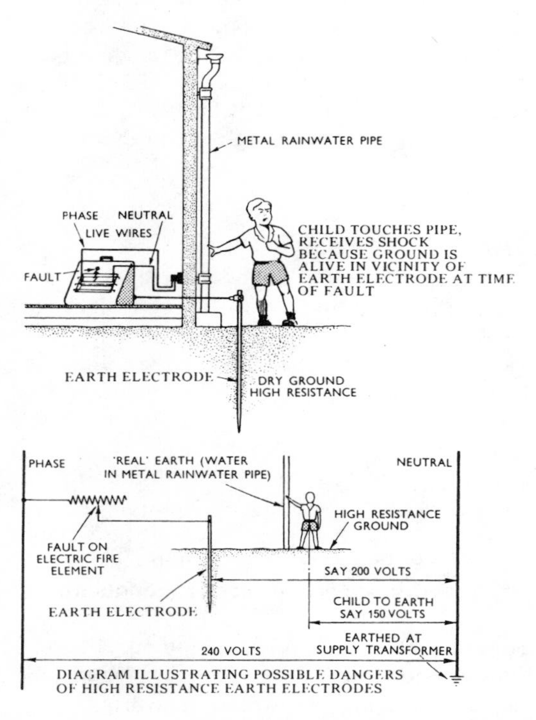

Figure 3.20. If the earth electrode is in dry ground having a high resistance, it is possible that a child or other person touching the metal rainwater pipe through which water is flowing to earth might receive a shock when there is a fault on an electrical appliance, unless measures such as the installation of earth leakage circuit-breakers prevent this danger from arising

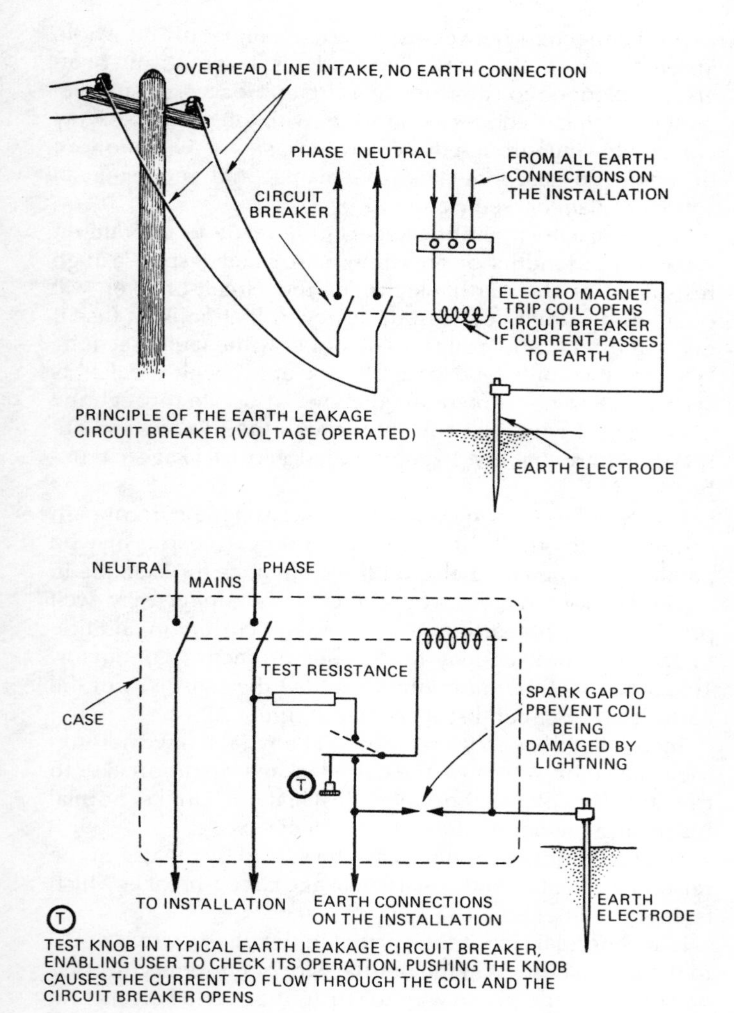

Figure 3.21. The principle of the earth leakage circuit-breaker (voltage-operated type)

The connection between the earth terminal of the whole installation and the earth electrode is taken through an electromagnetic coil attached to a circuit-breaker, or automatic switch, that is connected in series with the main fuse. Any current flowing through this coil will cause the switch to trip, or open, so disconnecting the mains, and removing all source of danger (see *Figure* 3.21).

Earth leakage circuit-breakers can be made to operate on several thousandths of an ampere, so that despite a high resistance at the earth electrode, the circuit-breaker will operate. It should be realised, however, that unlike a fuse it may cut off the whole supply and not only the faulty section.

More than one type of earth leakage circuit-breaker is available. Some – as mentioned above – operate through the passage of the leakage current through the trip coil and are referred to as 'fault-voltage-operated earth leakage circuit-breakers'.

This type has certain disadvantages. First, the current path through which the fault current must flow, to earth, may be paralleled somewhere else on the system, as for example in an immersion heater, where the water piping may well provide a second earth path for the fault current in addition to the path provided by the proper protective conductor. This may affect (but not entirely nullify) the sensitivity of the earth leakage circuit-breakers (see *Figure* 3.22).

Secondly, it has to be realised that any fault anywhere on the installation will cause the earth leakage circuit-breaker to trip and so cut off the whole installation. Unlike normal fusing arrangements, there is no selectivity.

The first of these problems can be solved by the use of the residual current operated earth leakage circuit-breaker which is described below.

In any normal circuit the current in the phase wire is equal to the return current in the neutral wire but if some current is passing from the phase wire to earth, these two currents no longer balance.

The residual current ELCB uses this condition, by employing in effect two windings, one carrying the phase current and the other the neutral, and so arranging the mechanical parts that if the forces exerted by these two coils become unequal, the circuit-breaker will trip (see *Figure* 3.23).

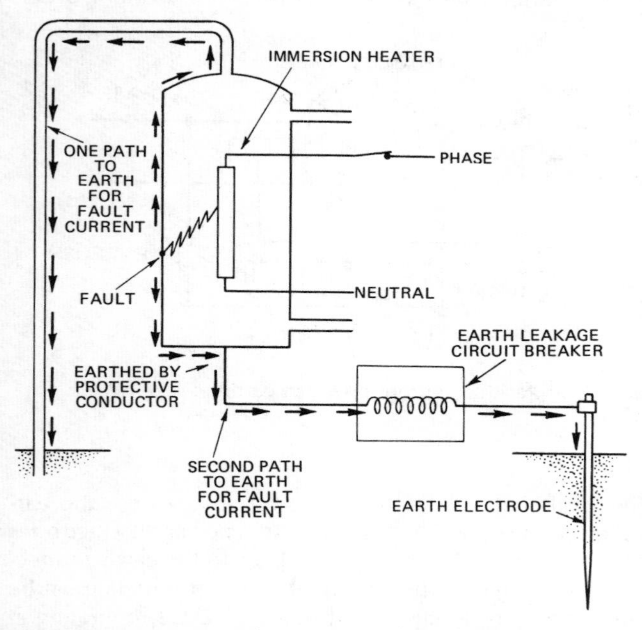

Figure 3.22. How a voltage operated earth leakage circuit-breaker may become inoperative if the faulty appliance has a parallel path to earth for the fault current, for example through water piping

The second problem, the lack of selectivity, can be solved only by splitting the installation up into sections that are entirely separate electrically, each being fitted with a separate ELCB, whatever type may be used. The circuit-breakers all have a common earthing point, but on the installation side of the voltage-operated type the protective conductors must be kept completely separate from each other. If this separation is carried out, the occurrence of a fault will result in only one circuit-breaker tripping, leaving the remainder of the installation in service.

All ELCBs are provided with a test knob, so that the user can make periodic checks that the device is in working order.

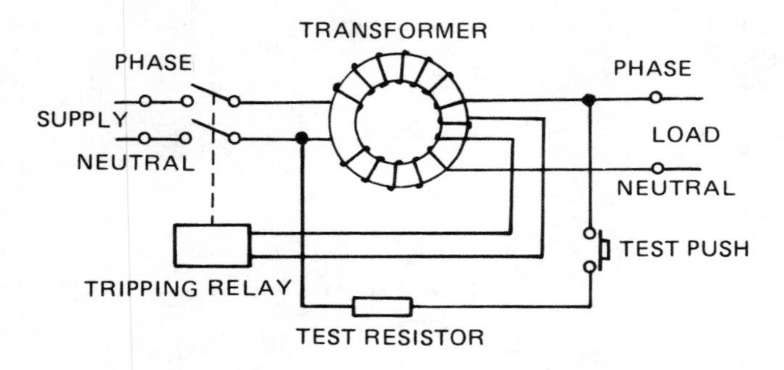

Figure 3.23. Residual-current-operated earth-leakage circuit breaker

The IEE Regulation (471-13) requires that, where the earthing on a domestic installation consists of earth electrodes only, with no electrical connection back to the earth connection at the source of the supply, the socket-outlets must be protected by a residual-current-operated ELCB, operating at not more than 30 mA. Also IEE Regulation 471-12 requires that a circuit rated at 32 A or less which is intended to supply equipment outdoors by means of flexible cords must be protected by a similar ELCB.

The voltage-operated type of ELCB is now hardly ever used on domestic installations, the residual-current type being the most common. IEE Regulation 413-2 states a preference for the residual-current type.

Protective multiple earthing (PME)

This system was originally used only in areas where difficulty was met in obtaining satisfactory earthing, particularly when the supply was overhead. It is, however, now being more widely applied by electricity boards, including urban areas where the supply is underground.

As we have seen earlier, the normal system of supply has the neutral conductor earthed at the supply transformer only, and it must not normally be earthed anywhere else since this could cause stray currents in such other services as telephone cables.

Sanction from the appropriate ministry is required before PME can be applied in a particular area. When the PME

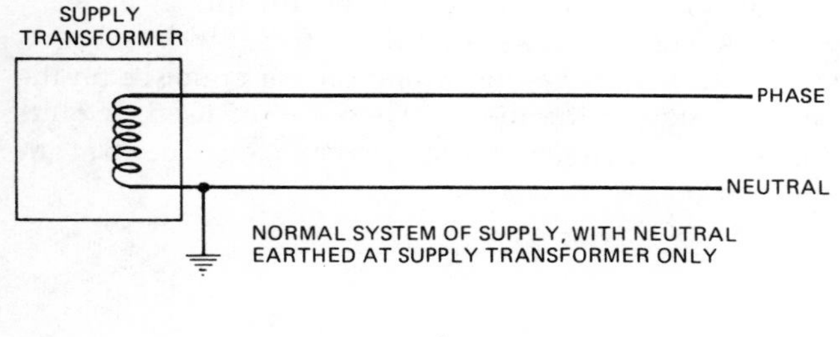

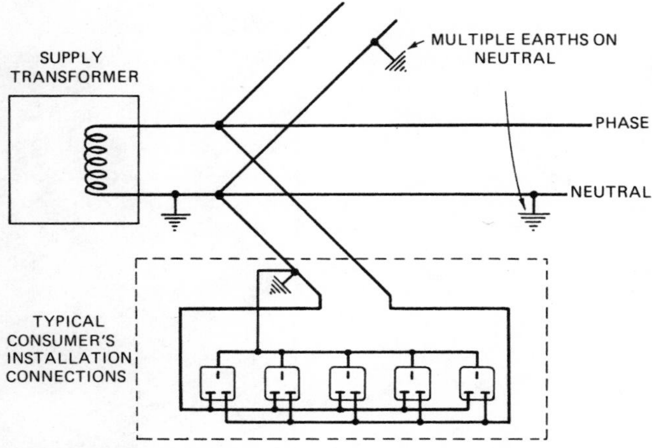

Figure 3.24. The principle of protective multiple earthing. The consumer's main earthing terminal is connected to the neutral by the electricity board at the meter position at each consumer

system is used, the neutral conductor is connected (by the electricity board) to the main earthing terminal at each consumer (see *Figure* 3.24).

When the PME system is applied, the neutral conductor on each installation becomes the earth point and all the protective conductors from the installation are connected to the neutral at the intake position. Thus a phase-to-earth fault becomes a phase-to-neutral fault.

When PME is employed, the bonding of other services (gas, water etc.) and exposed metalwork to the consumer's main earthing terminal is particularly important and the electricity board should be consulted on this at an early stage. IEE Regulation 413-2 and 547-2 refer.

Apart from the above, the wiring on the premises on the consumer's side of the meter, when PME is used, is quite normal and the neutral must not be connected to earth at any other point on the installation.

4

Physical plan of the installation

The layout plan

Position of intake point

An important factor to consider is the situation of the intake point (which must incidentally be agreed with the electricity board). One place to avoid is the cupboard under the stairs, which is likely to get filled with things like pushchairs and tennis racquets, so that (1) the meter reader has great difficulty in getting to the meter, and (2) if a fuse blows, confusion in the resulting darkness may become even greater as the householder first has to force his way through old trunks and then fumble around with fusewire.

It should be realised that the area electricity boards only allow a certain distance of 'free' cable, from the mains in the street, or from the pole line, to the supply point. This is often of the order of 10 metres. If a supply point position is chosen so that this distance is exceeded, the consumer must be prepared to pay for the extra cable involved.

A garage is often found to be a convenient place for the intake point, but care must be taken (1) to situate the fuses, etc., in a position agreed with the area electricity board, and (2) to allow ample room for cabling runs to the rest of the house, *and* for extensions to the consumer unit assembly.

Many new houses are fitted with a cloakroom or lavatory adjacent to the hall, and this is another place where the intake point may conveniently be situated, perhaps in a cupboard. A new development is the placing of meters on the outside wall of the premises in a special lockable cabinet, so that the meter reader can take the readings even when the household is unoccupied.

How many socket-outlets are needed?
The number of socket-outlets to be installed, and their positions, must be a matter for the owner of the installation, but the electrician should try and persuade him to be as generous as possible with outlets. In every household the usage of electrical appliances is always increasing.

A reasonable number of socket-outlets for a three-bedroomed semi-detached house is indicated in *Figure* 3.11.

In general socket-outlets should be installed in living rooms at about 500 mm from the floor level, to obviate the need to stoop down to insert the plug. In the kitchen, socket-outlets may be at table-top level, for convenience in using irons, food mixers, kettles and the like.

It is desirable to install at least one socket-outlet on the landing and another in the hall, so that vacuum-cleaner connections may be made without difficulty.

It should always be remembered that much time and material can be saved by careful planning and measuring-up before commencing work. The frenzied rush to the electrical shop for one more reel of cable, just as it closes, is a sign of bad planning.

There is another advantage in taking great care to prepare a drawn-out plan of the installation. It is almost inevitable that extensions will be needed some time in the future, and it is of course not impossible that a fault will occur in some part of the wiring. If the original wiring plan is available, both the work of extension and that of tracing and rectifying a fault will be greatly simplified.

Identifying tags attached to cables under floors and behind consumer units will be found useful if work is to be carried out in the future.

Size of cable

Having decided on the circuit layout, the next problem relates to the physical installation of the wiring.

The first point to decide is the size and type of the cabling.

The size of cable to be used is determined by several factors.

Lighting circuits

The actual loads should be used, when known, but each lighting point should be assumed to take at least 100 W. It is usual for domestic lighting circuits to be fused at 5 A, using 1.00 mm^2 PVC-sheathed twin and earth cable. Thus up to 12 points could be connected on one circuit (1200 W is equivalent to 5 A). However it is preferable to restrict the number of points to 8–10, the usual practice being to provide one lighting circuit for each floor.

Permanently connected appliances such as cookers, water heaters and radiators

In deciding on the circuit arrangements it is first necessary to estimate the maximum loads for the various appliances. The actual loads in practice are not necessarily the same as the connected, or highest possible, loads. Appendix 4 of the IEE Regulations for Electrical Installations gives information on the maximum demands applicable to commonly used equipment and on the allowances to be made for diversity. ['Diversity' allows for the fact that not all the parts of an appliance, or individual appliances, are likely to be in use at the same time.]

Taking a domestic cooker as a typical example we will assume a total possible loading with all items switched on to be 14 kW which is 58.24 A. However this situation is extremely unlikely to arise in practice and, applying IEE Appendix 4, the load is calculated as follows:

1st 10 A at 100%	10 A
30% of remaining 48.24 A	14.47 A
	24.47 A
Allowance for socket-outlet on cooker control unit, when applicable	5.0 A
	29.47 A

A 30 A circuit would therefore suffice but it might be preferable to provide a 45 A circuit so as to allow for a possible larger cooker in the future. A cooker should always be on a separate circuit from the consumer unit.

Ring circuit

The ring circuit is permitted, under the Regulations, only for use with 13 A socket-outlets fitted with fused plugs. The number of socket-outlets that may be connected has been set out earlier in Chapter 3.

Mineral-insulated cables

Mineral-insulated copper-sheathed cables employ single wires, not stranded conductors. These cables have higher current ratings than PVC-insulated cables of the same conductor size, mainly due to the higher permissible operating temperatures for MICS cables. *Table* 4.1 shows the ratings of the sizes most commonly used.

Flexible cords

Flexible cords should always be used as little as possible and should be kept as short as possible. They should never be used for permanent wiring.

Four types of flexible cord are commonly used:

for light-current work
for heavy-current work
for portable tools, etc.
for the connection of hot appliances.

For light-current work such as suspending lamps and connecting up lighting fittings and radio sets, where the current does not exceed 3 A, the type of flexible cord generally used is twin, plastic-insulated cord, with 0.5 mm^2 conductors.

For heavy-current appliances such as radiators, three-core plastic-insulated cord, sheathed or covered with braiding, is used, with 1.5 mm^2 conductors.

Table 4.1 Current rating of mineral-insulated cables exposed to touch or having overall PVC sheath

Nominal cross-section area of conductor (mm^2)	*Two single-core cables* (A)	*Twin cables* (A)
1.0	22	17
1.5	27	22
2.5	36	29
4.0	46	38

Based on information contained in the *Regulations for Electrical Installations*, 15th edition.

Table 4.2 Flexible cords

Size (mm^2)	*Current ratings* (A)	*Maximum permissible weight that a twin cable should support* (kg)
0.5	3	2
0.75	6	3
1.0	10	5
1.5	15	5

Based on information contained in the *Regulations for Electrical Installations*, 15th edition.

For portable tools, a three-core, plastic-insulated cable, with 1.5 mm^2 conductors, is suitable.

For connections to such appliances as irons, water heaters and any other appliance that can become hot, a butyl rubber or silicone rubber cable, of three-core construction, with special heat-resisting qualities, and conductors of 1.5 mm^2 size, is used.

Table 4.3 Current rating of single-core PVC insulated cables bunched and enclosed in conduit (see note)

Size of cable (mm^2)	*Two cables* (A)
1.0	14
1.5	17
2.5	24
4	32
6	41
10	55
16	74

Based on information contained in the *Regulations for Electrical Installations*, 15th edition.

Table 4.4 Current rating of twin and multi-core sheathed PVC insulated cables clipped direct to surface (see note)

Size of cable (mm^2)	*One twin core cable* (A)
1.0	16
1.5	20
2.5	28
4	36
6	46
10	64

Based on information contained in the *Regulations for Electrical Installations*, 15th edition.

Note: The ratings in *Tables* 4.3 and 4.4 are based on the cables being protected by cartridge fuses. If rewireable fuses are used the ratings should be multiplied by 0.725. The cables must be derated (allotted a lower current-carrying capacity) if the ambient air temperature is over 30°C. Conversely the ratings may be increased if the ambient temperature is below

30°C. Derating is also necessary in certain other circumstances and Appendix 9 of the IEE Regulations should be consulted when necessary.

Wire tables
Table 4.2 shows the current-carrying capacities and weight-supporting loadings applicable to commonly used flexible cables used in domestic and small industrial installations.

Type of cabling

The conduit system
If the heavy gauge steel conduit system is adopted, the system must be electrically continuous throughout: that means that all sections of conduit (piping) must be screwed together, with screwed sleeves to join each section of straight pipe, and properly screwed and terminated lengths running into metal junction boxes and the boxes that receive switches, ceiling roses, socket-outlets, clock connector boxes and all other fittings.

The conduit must be carefully measured before screwing, and after the thread has been cut with the proper dies, the ends must be reamered out to remove all sharp edges.

Steel conduit is made in various sizes, and *Table* 4.5 gives the mechanical characteristics of the sizes most commonly used in domestic installations.

It is laid down in the Regulations that all conduit installations must be completely erected before any wires are drawn in.

In deciding how many wires can be pulled into conduit systems, it must be realised that the friction caused by the pulling in of wires running close together can damage the PVC insulation.

The *Regulations for Electrical Installations* lay down the number of separate single-core cables that can be drawn into various sizes of conduit, and it is assumed that the reader who contemplates installing a system of this kind will have consulted the Regulations.

The method used to determine the cable capacities of conduits is described in Regulation 529-7 and Appendix 12. Each cable size is allocated a factor. The sum of all factors for

Table 4.5 Steel conduit sizes

Conduit external diameter (mm)	*Pitch* (mm)	*Tapping drill* (mm)	*Clearance drill* (mm)
16	1.5	14.5	18
20	1.5	18.5	22
25	1.5	23.5	27
32	1.5	30.5	34

the cables intended to be run in the same conduit is compared with a Table, and a suitable size of conduit is selected, this takes into account any bends or sets that are to be included in the installation.

Installing conduit

If conduit wiring is to be installed in a new house, the conduit must be placed in position at exactly the right time in relation to the other building work, or else there will be a great deal of expensive reinstatement of plaster work that has had to be cut into to run the conduit. This stage is usually called the carcassing stage.

When the the shell of the house is built, and the roof is on, but before the floorboards are laid and of course before plastering commences, the electrician has the ideal conditions for his work.

For extensions to existing installations it is usually necessary to chase out a trough in the plaster and bricks for the conduit to lie partly below the surface, otherwise, with a plaster depth of only 10 mm, the conduit would stand proud of the wall. For the outlet boxes, into which the actual socket-outlets are to be fitted, a deeper hole must be cut in the wall.

Right-angle elbows should be avoided wherever possible, because of the obstacle they provide to easy drawing-in of the wires. Bends are preferred, and these should have an internal radius of not less than 2.5 times the overall diameter of the conduit, e.g. overall diameter of conduit = 16 mm, minimum internal radius of bend = 16 × 2.5 = 40 mm.

Although a conduit system is in theory completely sealed, from end to end, air can gain access at the socket-outlets and switches, and moisture-carrying air brings with it the danger of condensation. This could mean that small quantities of water could run down into switches and other fittings, and give rise to corrosion. To prevent this, conduit systems should have drain holes situated at low points (in boxes).

Conduit runs should always be kept clear of gas pipes, water pipes, and any other wiring. This is because of the possible dangers of bimetallic corrosion through contact betwen different metals, and also because of possible sparking dangers, if a fault should occur on the electrical installation and fault current flows through the conduit itself.

As erection proceeds, draw wires should be inserted from junction box to junction box. These wires (or flat steel tapes) will greatly facilitate drawing-in. The Regulations state that all the conduit must be complete before any wires are drawn in.

When pulling in the cables, the best method is to bare all the ends of the conductors, twist them together and bend them over to form an eye and attach the end of the fish wire or tape to this eye. The procedure is shown in *Figure* 4.1. Care must be taken later to cut away the whole of the conductors used for this purpose, as they may have become damaged, before making them off into the fittings.

Always make certain that enough cable is pulled through. It is very unwise to leave only the bare minimum protruding from the box in the wall. If any slight slip is made during making off, there will be an extremely arduous business of pulling new wire in.

As a general rule, always try to avoid joints, even though they can be made in proper joint boxes, if there is no

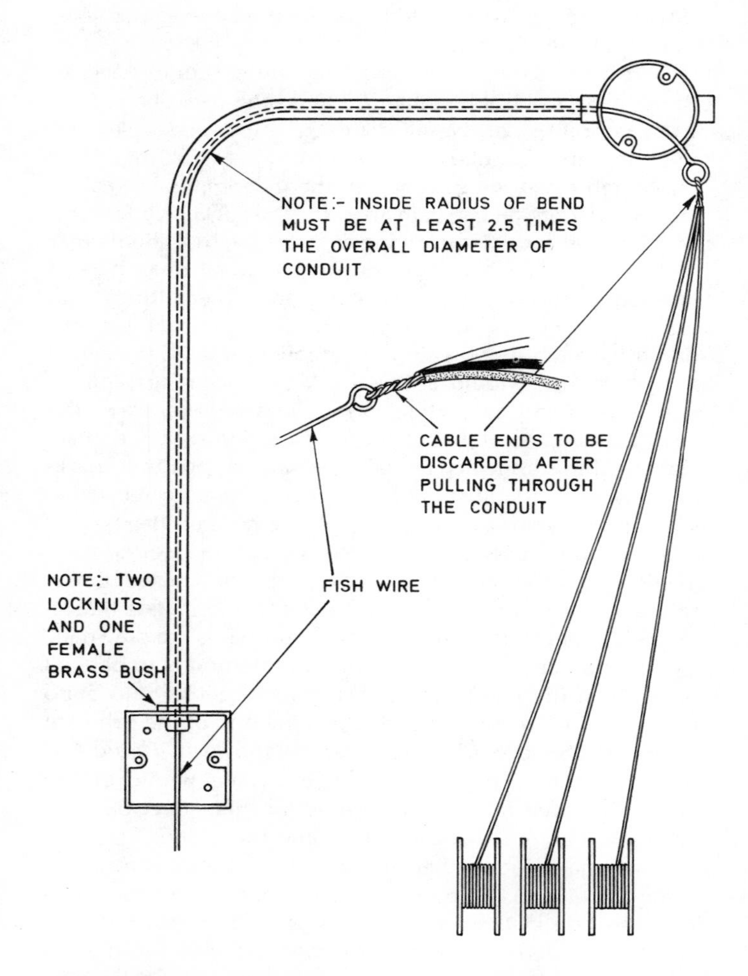

Figure 4.1. Use of fish wire and twisted cables when pulling into conduit. The twisted end must be discarded before connecting up, as it may become damaged

alternative. This means, of course, that short lengths of cable will get left on drums, and the electrician may think he has been wasteful. But he will have the satisfaction of knowing he has provided a sound job, and he will have avoided the laborious business of making joints in joint boxes which may – and often do – find themselves in difficult positions.

When it is necessary to pull in a number of cables, nasty kinks and knots may be avoided by 'combing' the cables through a piece of wood or stiff cardboard, with an appropriate number of holes, before entering them into the conduit. This will ensure that the cables are reasonably straight and will ease the drawing in process.

Sheathed wiring systems

PVC insulated and sheathed cables are by far the most commonly used materials in domestic wiring systems, whether for new installations or for extensions to existing wiring.

All wiring systems, as we have seen, must have proper earthing arrangements. With properly screwed metal conduit the conduit itself may act as the protective conductor, but where sheathed cable is used there is usually an uninsulated (bare) wire lying between the red and black wires within the sheath, to provide earth continuity.

The main points to watch when installing sheathed wiring systems are these:

(1) The greatest possible care must be taken to ensure that the cables are protected against mechanical damage.
(2) The cables have only a limited ability to resist heat, and must not be installed, for example, in the close vicinity of hot-water pipes or near flues.

If cabling has to be run in hot situations, the best choice is mineral-insulated copper-sheathed cable.

Preventing mechanical damage

Returning to the question of preventing mechanical damage, the installer must always try to visualise what *might* happen: for example, running cable in a trough cut into the plaster down a wall to reach a switch may be all right while the present occupier inhabits the room, but some future occupier might easily decide to fit a bookcase or a mirror on that part of the wall and drive nails or fixing screws right through the wire.

The best way of running sheathed wiring in plaster is to install a round or oval conduit for all runs, or at the least to chase out a trough for the wiring so that it can be covered with a metal channel. The use of conduit for runs to socket-outlets and switches has the advantage of enabling the cable to be pulled out if it has to be replaced, without breaking away the wall.

Alternatively, metal channelling fitted above the cable before making good the plaster will partly protect the cable, although complete protection is not ensured. A cautious person driving a nail or drilling for a wall plug would probably detect the metallic contact, if the channelling was in the line of fixing. The channelling will also protect the cable during plastering.

Similarly, if the floorboards are removed to run cabling beneath, it is rather tempting, when running at right angles to the joists, to cut a nick or groove in each joist, and run the cable across them in this way, afterwards replacing the floorboards. But some future occupier, not knowing the run of the cables, might well drive nails or screws into the floor, so penetrating the cable and even causing a fire (see *Figure* 4.2).

The proper method is to drill generously sized holes in the joist, at least 50 mm down each joist, and thread the cable through. 20 mm holes will allow three 2.5 mm^2 cables to pass through.

In general, if it is difficult to conceal the cables by running through confined spaces, or where thick stone walls are encountered, or if it is very difficult or impossible to obtain access below floors, it is better to run the cabling on the surface. People are not likely to damage a cable they can see. It is quite possible to run the usual cables in domestic premises so neatly that when painted to match the surroundings they become almost invisible, but are still apparent to anyone proposing to drill or drive nails.

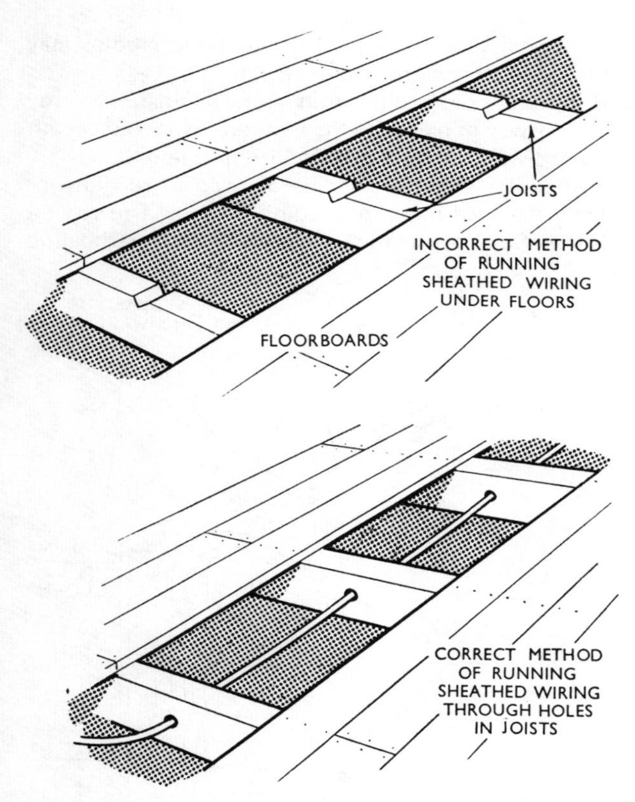

Figure 4.2. Running cable (or conduit) across joists: it is incorrect to run across the top of the joists in nicks, since nails driven through the floorboards could penetrate the pipe or sheathed cable and cause a fault. This method will also weaken the joists

Where the wiring need not be covered for aesthetic reasons, such as in a garage, it is better to run it in such a way that it can be clearly seen (*Figure* 4.3). By using this method damage to the wiring will be less likely than if it is indifferently concealed.

Cables must always be supported by some form of clasp, usually taking the form of plastic clips (*Figure* 4.4) or buckle clips, the normal spacing being 250 mm apart. Saddles may be used to embrace several cables (*Figure* 4.9).

Where a number of cables run *along* a joist, half-way down, an easy way of fixing is to use saddles from old sheath ends, secured with tacks of at least 15 mm in length.

Sheathed cables should never be allowed to run unsupported for a distance much exceeding 300 mm. The reason for this is that if they are at any time subjected to overheating

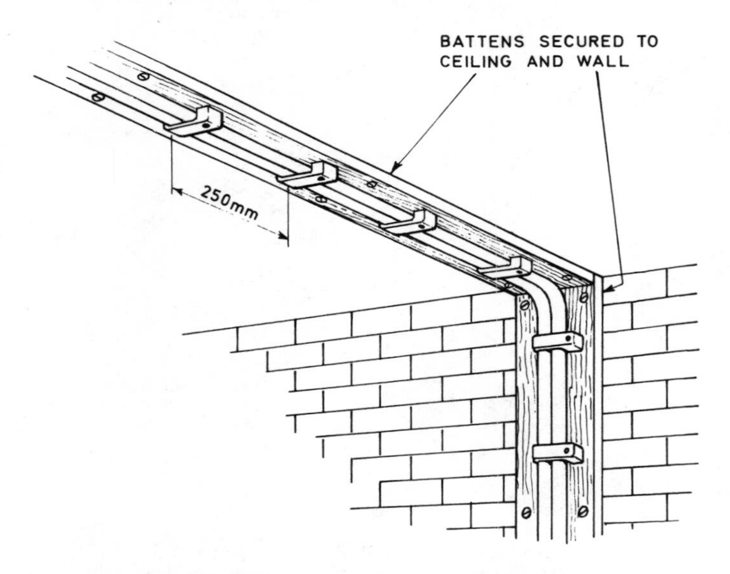

Figure 4.3. Sheathed wiring run bare on battens

– and this may be due not only to the passage of excess current, but also through the proximity of a flue, or a hot-water pipe – the sagging cable may become distorted so that the PVC insulation tends to drop away from the conductors, and might even leave them bare.

It is always bad practice to tie or tape sheathed cables to existing water or gas pipes – especially the latter. Wood battens should be run along the path to be taken by the

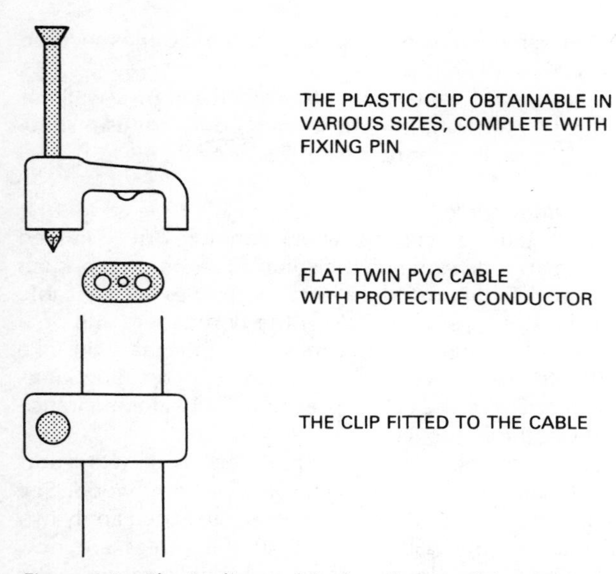

Figure 4.4. Plastic clip used for installation of sheathed wiring

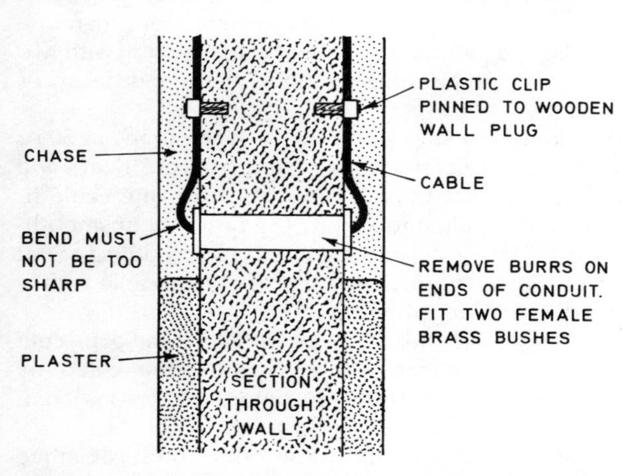

Figure 4.5. Running sheathed cable through walls: protective conduit should be employed, it need not be earthed

cables, well removed from the piping, and the cables secured by means of saddle clips.

Where sheathed cables have to pass through a wall, or through any partition, the best method is to use some lengths of protective conduit (see *Figure* 4.5).

Running out the cable

The best system to employ, when running out sheathed cables, is to arrange some sort of spindle for the drums, such as a broomstick supported on wooden frames. If the cable runs off the drum as it lies flat on the floor, there will be a danger of producing a twisted mess of cable that will take much time to disentangle and may in any case result in kinks and untidy twists in the cable, preventing a neat appearance in the finished job.

To secure the neatest job, experiment with two short lengths of cable, say 300 mm each, on a piece of wood. See how near to each other the cables will lie, and then apply two sets of plastic clips, 250 mm apart. If the cables are now clipped in, and lie as close together as possible, measure the spacing between the clips at each point, and adopt that measurement along the runs of cable where more than one circuit is to be laid. Nothing looks worse than wiring with two or three cables untidily running along at varying distances apart.

When running out sheathed cable, which comes in 50 or 100 m drums, great care must be taken to avoid twists and kinks, as mentioned earlier. These may harm the cable if, when kinked, it is pulled too tight. The best way of smoothing out the cable is to clamp a smooth round object, of at least 30 mm diameter, in a vice and pull the cable tightly round it, one hand opposing the other.

PVC sheathed cable tends to be stiff to handle in cold weather. This problem can be overcome to some extent by storing the reels in a warm room before operations commence.

Assuming the installation (or extension) is to use sheathed cable run beneath the plaster, the first process is to mark out, as accurately as possible, the cable runs.

The cable could be covered with steel channelling (although this is not essential), and this comes in standard sizes. A run commonly used is two 2.5 mm^2 cables, as employed on a ring circuit, and is about 45 mm wide overall. If this size of channelling is used, a chase, or trough, about 50 mm wide needs to be cut in the plaster.

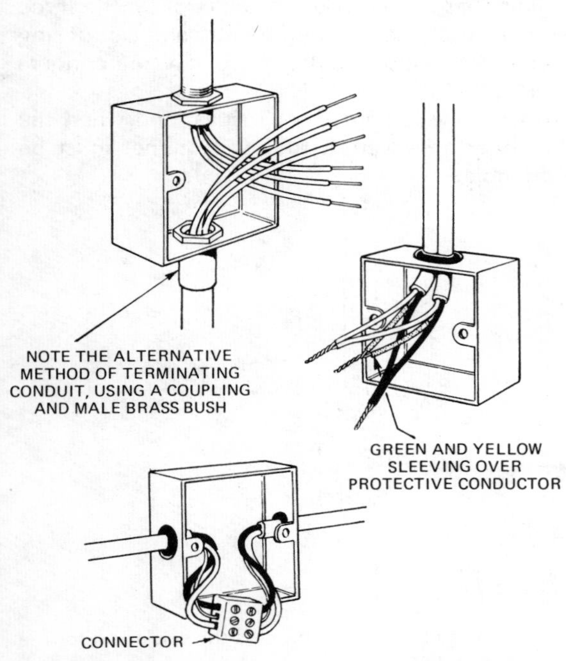

Figure 4.6. Types of box used both on conduit and sheathed wiring systems showing details and method of baring and twisting the conductors

The best way to cut the plaster without bringing away more surface than is required is to employ a wide flat chisel and cut diagonally inwards at both sides of the chase. The plaster can then be levered out with a knife.

The cable will terminate in a box, on which will be mounted the switch or socket-outlet.

Fitting steel boxes
Steel boxes (*Figure* 4.6) are made in different depths. The shallow boxes are 16 mm deep, and do not allow much room for more than one cable and the resultant interconnecting wiring. These are eminently suitable for switches. A deeper box is 47 mm deep, and allows ample room for PVC connectors, when a joint has to be made in a cable, or for the three sets of wires that result, say, from the entry and exit of ring circuit connections and also the departure of a spur connection.

These boxes have knockouts on all sides, and when the knockout has been removed, a rubber grommet must be inserted in the hole.

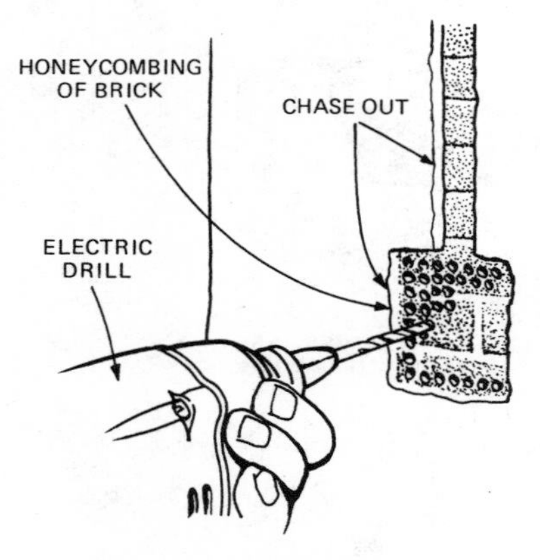

Figure 4.7. Using the electric drill to prepare the brickwork for the large hole needed for a deep box

The brickwork must be cut away to receive the deeper type of box. Over the whole area to be cut away (80 mm square for a one-way box, 80 mm by 160 mm for a two-way, and so on) holes are drilled in the brick with a size 10 or 12 masonry drill, and then a sharp chisel will soon clear away the honeycombed brick (see *Figure* 4.7).

The box must be sunk in so that its top edge is about 5 mm below the general plaster level of the wall.

Holes for screws will be found in the base of the box, two of them usually being oval to allow for adjustment. Holes must now be drilled in the wall (a size 8 woodscrew is usually used, so an appropriate drill or bit must be used), a wall plug inserted, and the box screwed firmly to the brick.

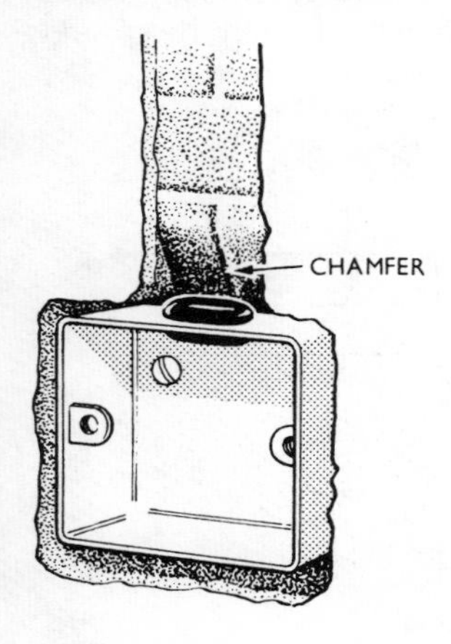

Figure 4.8. Chamfering brickwork to allow easy cable entry into box

At this point a small spirit-level is useful, because the levelling of the box will ensure that the front plate of the switch or socket-outlet will also come exactly level. If one switch-plate slopes one way and an adjacent switch slopes in the opposite direction, the result clearly indicates bad workmanship. The spirit-level may be used to level the box, using the tolerance provided by the oval holes, before the screws are finally tightened.

On any side of the box where cables are to enter, the brickwork should be chamfered away as shown in *Figure* 4.8, so that there is a smooth slope and not a sharp edge.

Fitting cables and channelling

The next step is to run the cables in the chase cut in the plaster. The cables are naturally springy, and it is not easy to get them to be flat ready to receive the metal channelling. It has been found worthwhile to plug the brick in one or two places (easily accomplished with an electric drill) and to fit a few clips along the run to hold the cables in position as shown in *Figure* 4.9.

The channelling may be secured by tacks or screws. Sometimes a satisfactory fixing (bearing in mind the plaster will soon hold the channelling firmly) may be secured by tacks driven into the crevices between bricks. But the springy

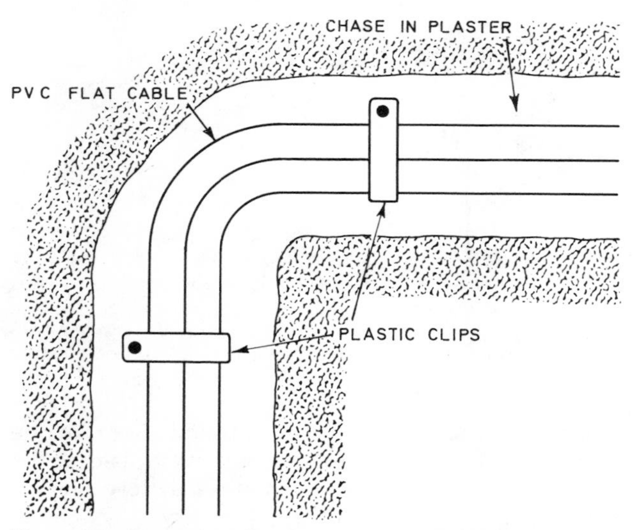

Figure 4.9. When installing metal channelling, keep the cable flat against the wall by using plastic clips under the channelling. *Note:* the radius of the sharpest bend must be at least four times the overall diameter of the cable (too tight a bend will damage the PVC). Also, all hidden cable should be run either vertically or horizontally, to give indication of run

nature of PVC-insulated cable, especially, will often mean that screws, driven into proper plugged holes, say every 300 mm along each edge of the channelling, will have to be employed to secure a neat job in which the channelling is properly sunk well below the plaster level.

The channelling should be cut so that it forms a continuous cover for the cable. This means that it comes right up to each box. Sharp edges should be filed so that there is no danger of subsequent pressure accidentally cutting through the insulation.

Making good

The cables cannot be made off into the boxes before the wall is 'made good' – plastered and painted. Otherwise the switches and other fittings could become covered in paint.

The next step is to make good the plaster. The best method is to use first a cement and sand mixture (half cement, half sand) to fill in the main part of the trough containing the cables and their protective channelling, right up to the normal plaster surface layer.

An important point to remember here is that if the cement is allowed to harden for too long, it will be difficult to rake away the top layer to allow for subsequent plastering to restore the smooth unbroken surface of the original wall. Therefore the cement layer should be given, say, not more than 2 days to set, and then the surface should be scraped so that it lies about 5 mm below the normal wall level. The final finishing coat of plaster can then be applied. Priming and final painting can follow when the plaster has fully dried out.

Making off the cable ends

Perhaps the most likely source of trouble in an electrical installation lies in bad workmanship when making off cable ends.

When baring the insulation from the conductors, savage tearing with a sharp knife may easily cut through or nick one or more strands of wire, and these loose pieces of wire may come adrift and cause short-circuits, or may reduce the current-carrying capacity at the terminal, and so cause overheating. The best practice is to use a cable stripping tool (*Figure* 4.10).

When bringing two (or more) conductors into a terminal, enough insulation should be bared from each to allow the wires to be twisted together, with only enough bare copper showing for proper insertion in the terminal. The wires may be twisted with flat-nosed pliers, and the twisted ends cut off neatly with side cutters.

Only practice and experience will show exactly how much length of single conductor should be left outside the cable sheath, and of this, how much should be bared to make the connection. If too much wire is bared, there is a danger that as the box is closed up, the wires may become twisted or compressed against each other so that short-circuits occur. If on the other hand too little 'tail' wire (wire outside the sheath) is made available, making off may be difficult, and this may lead to the physical impossibility of tightening all the terminals as they should be tightened.

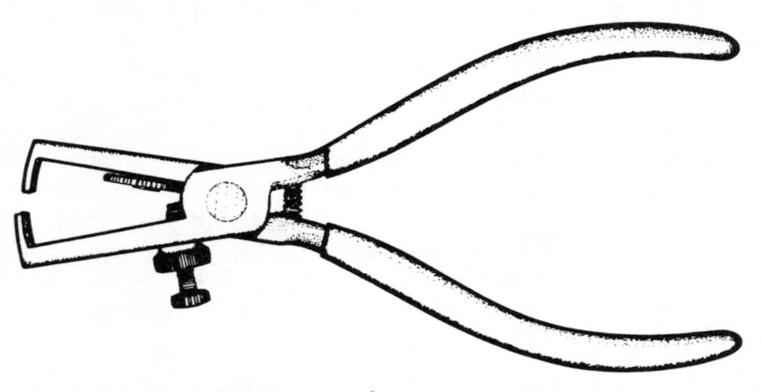

Figure 4.10. Cable stripping tool

Loosely tightened terminals can easily cause internal sparking and overheating. A high resistance is introduced into the circuit at this point, and insulation may become charred and dangerous, and springs in switch contacts may become overheated and lose their tension, and so provoke further trouble.

As each fitting – switch, ceiling rose, or socket-outlet – is completed, it should be *gently* pushed into position, the electrician taking every opportunity to make absolutely certain that as the fitting goes into its box none of the wires is being twisted too sharply, or compressed against a sharp edge, or brought into contact with the live terminals.

5

Accessories and fittings

The consumer unit

Starting at the consumer unit (*Figure* 5.1) where the installer's responsibility commences, this component should be chosen with considerable care with two viewpoints in mind:

(1) Ease of installation of the original wiring.
(2) Ease of adding new wiring when the occasion demands.

For an average domestic installation, the consumer unit might well include eight fuseways, allocated as follows:

two for the two ring circuits, 30 A each;
one for the cooker circuit, 30 A or 45 A;
one for the immersion heater, 15 A;
one for a 3 kW fixed radiator in the living room, 15 A;
two for lighting circuits, 5 A each;
one spare.

But there will be many variations on this.

As mentioned earlier in this book, careful studies concerning diversity of use, carried out over many years in many different types of household using various kinds of appliance, have shown that it is safe to allow for considerable diversity: in other words, to plan the wiring and fuses and other appliances on the basis that not all the potential, or possible, loading will occur at the same time.

For example, in the *Regulations for Electrical Installations*, for individual domestic installations, the *suggested* diversity factor that can be applied for example, to fixed heating and power appliances other than motors, cookers and water heaters, is 100% full load up to 10 A, plus 50% of any load in excess of 10 A.

Taking this suggestion as a guide, if in a four-bedroomed house there are 2 kW fixed electric fires in each bedroom, the total potential loading would be 4 × 2 = 8 kW = 33.28 A. But the diversity factor suggested means that the cable size could be proportioned for the first 10 A plus 50% of the remaining

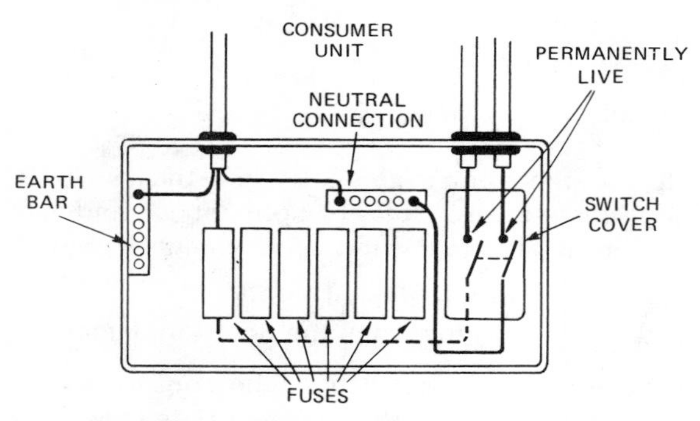

Figure 5.1. The interior of a typical consumer unit, showing phase (live) connection, neutral connection, earth bar and the cover over the switch, whose terminals remain alive even when the switch is off

current (33.28 − 10 = 23.28 A) or 11.64 A. In total, therefore, the current that should be allowed for is 10 + 11.64 = 21.64 A, instead of 33.28 A. This also applies to the fuses and the other fittings on the mains side of the installation but *not* to the final circuits.

However, it must be emphasised that this (and other suggested diversity factors) are only guides. The person responsible for the installation must make his own judgement.

For example, while for a cooker the same kind of diversity is suggested, if this cooker is installed, say, in a nursing home where special meals are prepared, or in a large house where a large family lives, or perhaps in a small café, then diversity does not necessarily apply.

The worst that can happen if too *large* a cable and consumer unit are installed is some small, *apparently* unnecessary expense at the time of installation.

The worst that can happen if too *small* a cable and consumer unit are installed is overheating (with possible fire danger), frequent blowing of fuses, and a probable need to embark on a costly and inconvenient programme of rewiring.

Each consumer unit will have its own main switch, fuseholders for each way, and neutral terminals and earth terminals.

There are metalclad and insulated enclosures for consumer units. If a conduit system is being installed, a metalclad type will be used. For sheathed cable systems, the insulated enclosure is more usual, but the metalclad type may be used if desired, care of course being taken to ensure that the metal enclosure is properly earthed.

All types are provided with knockouts. These are either sheet-metal plugs, pressed into holes appropriate for conduit or wire entry, or else (in plastic boxes) clearly defined thin places in the plastic case, which can be tapped out by a careful tap with a hammer.

The Regulations state that all enclosures surrounding live parts must be completely closed. This is required to prevent the ingress of moisture, insects, and dust. So if any knockouts are not required for wiring they must be left intact, or if already knocked out, closed by means of rubber plugs.

Where conduit enters a consumer unit it must be fitted with a smooth brass bush at its end, to prevent the wires being damaged on the rough edges of the pipe. When bringing sheathed wiring in, close-fitting rubber grommets must always be used. These grommets have the dual purpose of protecting the wiring from damage when passing through the rough edges of the hole, and also blocking the hole around the wiring to seal the interior of the box.

Consumer units may be installed straight on to a brick or concrete wall, but there are several disadvantages if this method is used. First, brick walls may be damp, and this will ultimately cause rusting, even on the best galvanised boxes. Secondly, to obtain the neatest wiring layout, particularly with sheathed wiring, it is desirable to bring the wiring in from the back, and this is not possible if the boxes are flush against the wall (for conduit wiring, top entry is usually more convenient).

The ideal method is to make up (or obtain ready-made) a strong wooden board, of not less than 13 mm timber, supported on battens at either side so that it stands about 50 mm

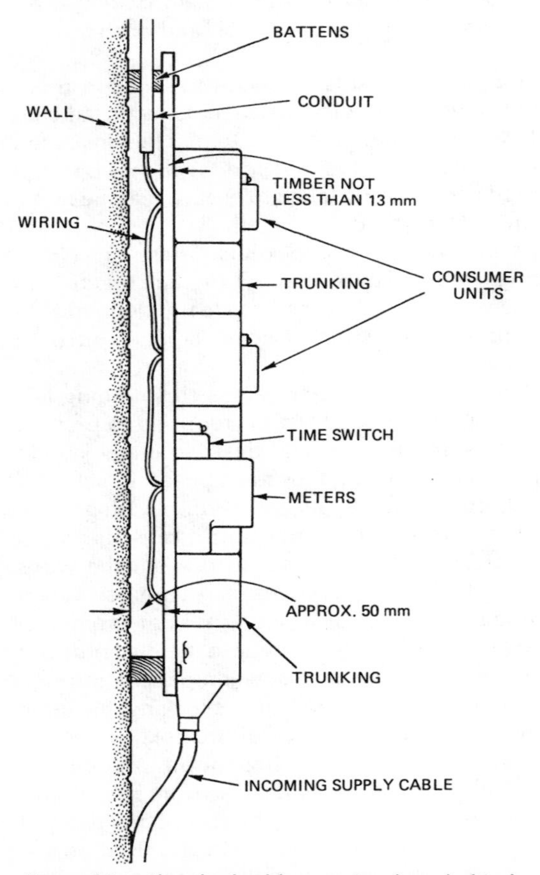

Figure 5.2. Sheathed cable entering from behind consumer unit

away from the wall (see *Figure* 5.2). This board is then firmly fixed to the wall at its left and right edges. The sheathed wiring, brought in from above on battens, can then be taken into the consumer unit through holes of generous size (not less than 20 mm diameter) drilled in the board at points carefully measured out to coincide with the back knockouts.

With ample space behind the wooden board, the wiring will not be unduly pinched up, so avoiding the possibility of damage and overheating. Moreover, new wiring can easily be added. Further holes can be drilled in the baseboard by removing the fixing screws of any one consumer unit, and – if proper slack has been left in the wiring behind the board – slightly moving it out of place so that the hole can be drilled and the new wire brought through.

For larger installations, the area electricity board often provide a consumer's termination unit in which the supply cable sealing box, the main fuses, and the meter are mounted in such a way that a trunking connection may be made directly, with the unit mounted, as it were, on top of the trunking, see *Figure* 1.7.

It should always be remembered that space should be provided on the baseboard for possible additional consumer units. For example, if off-peak storage heating is to be added, not only will an extra unit be needed, but a time-switch and possibly a contactor will have to be installed.

Again, if the size of the installation grows so that a 3-phase supply mains has to be brought in, there will be the need for separate consumer units for each phase; and if this move has been necessitated by the installation of off-peak storage heating, there will have to be units for the red, yellow and blue phases on-peak, and for the red, yellow and blue phases off-peak, making six in all.

Wiring accessories

In running out the wiring from the consumer unit, one cardinal principle must always be observed. There should be no joints in the wiring runs other than those made at proper fittings, using screw-down terminals of suitable size.

Care must always be taken to avoid what is unfortunately a common device for fixing more wires into a terminal than it will properly hold, or wires larger than those for which it was designed. Many amateurs have been known to cut away several of the bared strands of the conductors, to make them fit into the terminal. *This practice is entirely wrong, and may well be highly dangerous.* The reduced copper section at this one point may well give rise to considerable overheating, without blowing any fuse, and it could therefore happen that a fire is caused through a switch or socket-outlet becoming overheated.

Joint boxes

With sheathed wiring systems, round joint boxes are available. The electrician must be certain that the terminals are large enough to contain the required number of wires of the size to be used.

It should be mentioned, however, that many of these boxes are only suitable for use with the smaller sizes of cables, usually less than the 2.5 mm^2 size. Some boxes are specially made to take three 2.5 mm^2 conductors, for example where a tee-off is required.

The best practice is to avoid boxes wherever possible. Lighting circuits may be looped in, using four-plate ceiling roses, to avoid the necessity for joints (as mentioned later in this chapter) and, on ring circuits, any spur connections ought if possible to be made at socket-outlets, since the terminals on these units are capable of carrying three 2.5 mm^2 conductors.

Joint boxes are designed with knockout sections, and these should be broken out in such a way that the cable fits snugly and does not allow empty spaces around it where dust and insects may penetrate.

Where there is considerable likelihood that moisture or fumes or dust will penetrate, the proper box to use is a conduit type with sealing glands (that can be bought from the cable suppliers) so that the cable is properly sealed in to the box entry holes.

Metal boxes for switches, socket-outlets, etc., should be used with sheathed wiring systems, although plastic-type boxes are available. The protective conductor brought in with the cable must make proper connection with the box itself. An earthing screw is sometimes provided for this purpose, but the socket-outlet or other device usually has an internal arrangement whereby the earth connection terminal on the socket-outlet body has a metal strap connecting it to the screws used to fix the device on to the box, thus providing an earth connection when the socket-outlet is assembled.

A short cut sometimes adopted by amateurs, where they wish to earth a metal box, is to trap the earth wire beneath the lid of the box, holding it tight by means of the fixing screws. This means that a gap is left, through which moisture may penetrate. This short cut should not be employed.

Where plastic boxes are used, care must be taken to see that holes are not left in such a way that subsequent plastering will fill the box with liquid plaster.

Whatever type of box is used, it should be chosen so that there is ample room for all the wiring, and that when the switch or other fitting is applied, the wiring will not be squeezed up so tight that there is danger of a sharp point on the back of the switch pressing the wires against the metal part, with the possibility of ultimate breakdown of the insulation.

There is another danger that must be carefully avoided, and that is of cutting off the ends of the cable a fraction too short, so that when the switch is pressed home and the retaining screws are tightened, one or other of the connections is being gradually strained either to breaking point or is pulled out of the terminal.

Only experiment and practice on a dummy circuit, using some scrap cable, will show how much cable should be left.

All fittings – boxes, surface switches, socket-outlets – must allow for the outer sheath of the cable, in sheathed cable systems, to be brought well inside the protection of the box – in other words, the unprotected cores of the cable, even with their normal insulation, must never be accessible outside an area protected by some form of box.

Ceiling roses and connectors

Ceiling roses, for pendant lamps, are plastic with non-flammable back plates, called pattresses.

These ceiling roses often have four terminals instead of the three that might be expected. This is because many electricians use the loop-in system (*Figure* 5.3), which uses a little more wire but involves less labour than the alternative.

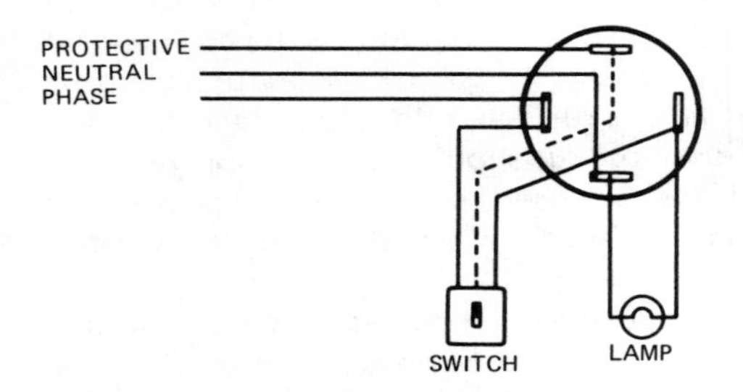

Figure 5.3. The loop-in system for ceiling rose wiring

The loop-in system has the advantage of not requiring a joint box as was necessary with older systems (see *Figure* 5.3). If, however, there is some good reason for using the joint box, care should be taken to ensure that no bare conductors are exposed. The conductors must 'fill' the terminals, which may require the conductors to be doubled back on themselves and that the terminal screws are tight (*Figure* 5.4).

Connection of fixed appliances

Most fixed appliances, with sheathed cable systems, are connected via a fused spur to the ring.

Other appliances such as fluorescent lamp fittings and towel rails are generally provided with bases of such a type that either they will fit directly on to the standard galvanised

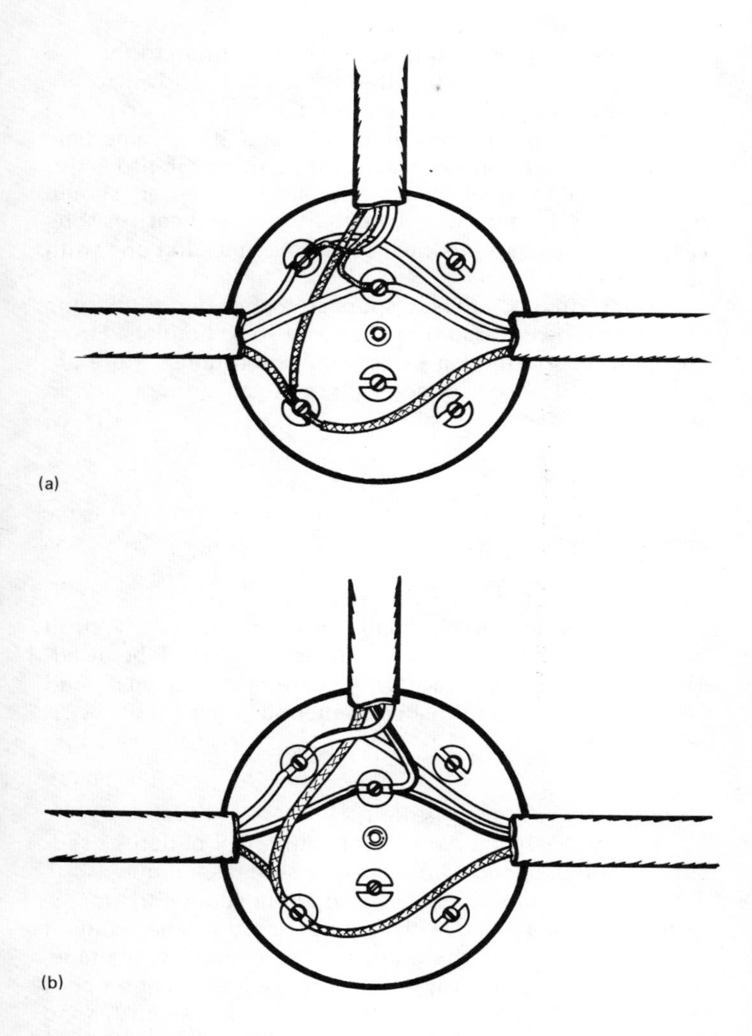

Figure 5.4.

(a) Joint box showing poor workmanship – excessive insulation removed, and sleeving omitted from protective conductor

(b) Correct termination with green/yellow sleeve over protective conductor

steel conduit termination box, or else they fit on to the wall or ceiling in such a way that they form a complete enclosure over the properly bushed end of the conduit, or of the protected end of the sheathed cable, *and* at the same time are provided with proper means for securing the protective conductor connections. Fixed floor standing heaters should be connected via a short length of flexible heat resisting cable from an adjacent switch or fused connection unit, with switch.

On ring circuits, unfused spurs are connected to the ring via junction boxes or into the backs of socket outlets. Fused spurs are connected via fused connection units; these are available in flush or surface type, and can be mounted in standard conduit boxes.

Special fittings

Clock connectors

For clocks, where the current consumption is very small indeed, special fused clock connectors should be used. These moulded plastic accessories are neat and small, and are often connected into a convenient lighting circuit. A 2A fuse is incorporated.

Outlets for fixed storage heaters

For storage heaters a switch unit with a flex outlet is used, and may be obtained with a flush or surface mounting.

A special problem may arise in connection with storage heaters that are fitted with fans to boost the heat output when needed. Since the boost may be needed at any time, these fans must obviously operate from a normal, or on-peak circuit. Special fittings are available, whereby the heater and fan can be connected to a single unit, designed to allow two circuits to be used, isolated from each other (*Figure* 5.5).

When the premises are fed by more than one phase of the supply, the two feeds to the switch unit should be on the same phase if possible. If this cannot be achieved, a warning notice (see IEE Reg. 514–4) and phase barrier should be fitted in the switch unit; these are readily available.

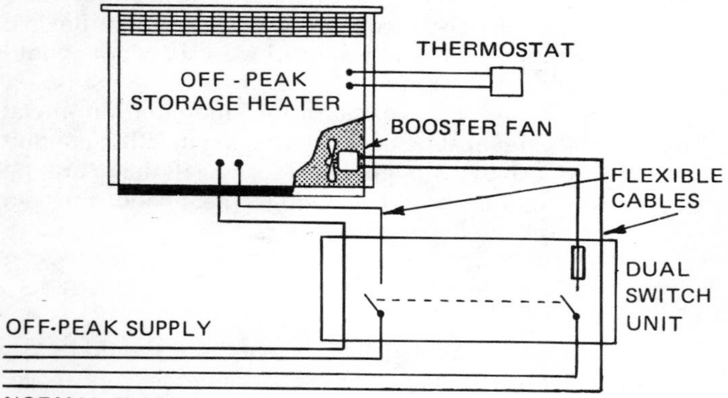

Figure 5.5. Connection diagram of storage heater with fan. The two switches should be mechanically interlinked, so that the heater is completely isolated in one operation

Floor-mounted socket-outlets

Many manufacturers provide a special type of floor socket-outlet, fitted with a strong cover that may be screwed into place when the socket is not in use, so that no dust or moisture can find its way into the fitting, and furniture can be confidently allowed to stand on it. Such a special fitting should always be used when the circumstances make it necessary.

Outdoor fittings

To prevent ingress of rain, and condensation inside, specially designed weatherproof switches, socket-outlets, lamp-holders and all other fittings should always be used outdoors.

Galvanised steel conduit boxes should be used, even if conduit is not being employed, since sheathed cable can be brought into these boxes, in a watertight fashion, by means of the special glands supplied by the manufacturers. If there is any doubt about watertightness, most cable manufacturers supply a type of plastic compound that remains flexible indefinitely to plug the intake ends of the cable coming into outdoor-mounted fittings.

Switches should be weatherproof, mounted in metal boxes. Great care must be taken in bringing in either conduit or sheathed cable (with glands) to see (1) that earthing continuity is carefully preserved, and (2) that no moisture can enter, whatever the direction of the rain.

Bathrooms

It is contrary to the IEE Regulations that there should be any portable electrical appliance or socket-outlet (except shaver sockets to BS 3052) in a room containing a fixed bath or shower. Switches in such rooms, except cord-operated types, must be normally inaccessible to a person using the bath or shower. This does not preclude the use of wall-mounted switches in suitable circumstances. Lampholders used in bathrooms must be of the all-insulated type, and must be fitted with a protective skirt, as illustrated in *Figure* 5.6.

The two most commonly used special fittings for bathrooms are the shaver socket and the ceiling switch.

Shaver sockets

To ensure that no user of an electric shaver could receive a shock while shaving in wet conditions, specially designed shaver sockets must be used in bathrooms.

Shaver sockets use a small transformer, the secondary side of which – the side that supplies the shaver itself – is entirely isolated from earth, so that no one touching any part of this circuit could receive an electric shock, as there would be no return path to earth.

Only shaver sockets which comply with British Standard 3052 (which lays down the standard of insulation between the primary and secondary windings of the isolation transformer, and specifies other safety precautions) may be used in

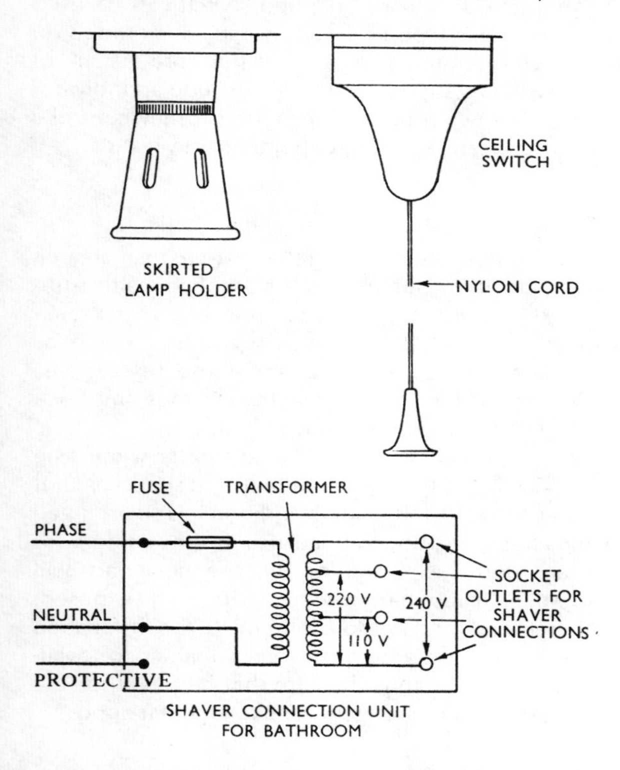

Figure 5.6. Bathroom fittings. Comprising special shaver socket, lampholder with skirt, and ceiling switch

bathrooms, and they must be properly earthed. Shaver sockets (without a transformer) are available for use in rooms other than bathrooms, and these should comply with BS 4573.

Cord operated switches

Where a wall-mounted switch is not permissible in a bathroom, one way of complying with the IEE Regulations is to place the switch outside the room, and this practice is often adopted. Alternatively a cord-operated switch can be used inside the room. This type of switch is usually mounted on the ceiling and is operated by means of a pull cord, usually of nylon. Such switches should preferably include an indicator light, which glows when the switch is 'on', if they control a non-luminous appliance, such as an immersion heater.

Apparatus installed in flammable atmospheres

If electrical apparatus is to be brought anywhere near an area where petrol, paraffin, butane gas, or any flammable substance is used, or where the vapour that rises from such substances may persist, then various regulations made by local authorities must be obeyed, and these regulations, added to the *Regulations for Electrical Installations*, impose a very high standard of practice in such situations.

To start with, specially designed flameproof gear must be used. It is rigidly tested to ensure not only that sparks cannot escape and cause explosions, but in addition it is so arranged that if flammable gases are trapped inside the conduits or boxes, and an internal explosion occurs, the metal parts will resist it and will not cause bare electrical wires to be exposed.

The beginner is advised not to contemplate the installation of flameproof gear unless he has professional advice available, and in addition he must be sure that he has studied – and complied with – the stringent regulations that apply.

Bell transformers

Electrical bells and chimes are often supplied at 6 or 12 V by means of small transformers, fed from the mains.

Such transformers are usually supplied from a convenient lighting circuit, and may be fused at 2 A; some bell transformers have their own fuses inside the case. As we shall see in Chapter 8, the extra-low-voltage wiring from the transformer to the bell and pushes must be kept quite separate from mains-voltage wiring unless the ELV wiring is insulated for mains voltage. IEE Regulation 525-3 refers. Auto-transformers (mentioned under 'Transformers', p. 142) must not be employed.

Double-pole switches

We have seen earlier that with the phase and neutral supply system the switches used (other than the main switches) are normally of the single-pole type – that is, they break the phase conductor only, and not the neutral.

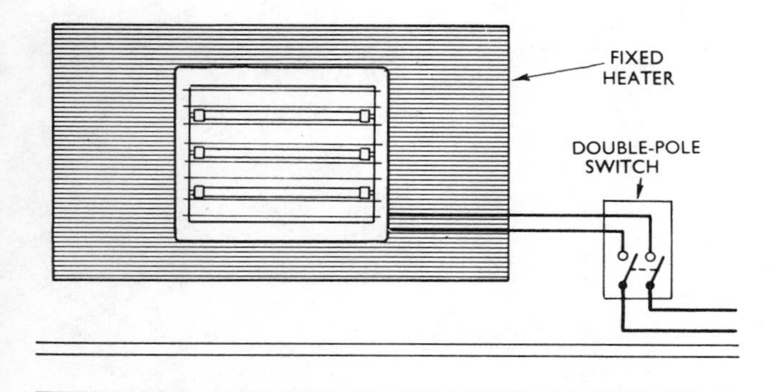

Figure 5.7. Double-pole switch used for heating appliances where elements can possibly be touched

But there are exceptions to this rule. IEE Reg. 476–17 specifies that double-pole switches shall be used for any *fixed* heating appliance *where the heating elements can be touched,* i.e. both conductors feeding the appliance must be interrupted by a switch placed near the appliance (see *Figure* 5.7). (This would apply for example, to an electric fire permanently installed on a wall, where the elements could be touched by inquiring childish fingers when the fire appears to be safe as it is not glowing.) For the purpose of this requirement the sheath of silica-glass-sheathed element is regarded as part of the element.

Double-pole switches and fused connection units are readily available, and will mostly fit into the same boxes (or occupy the same space as surface units) as their single-pole counterparts.

6

Extensions to old installations and temporary wiring

Extensions to old installations

Extensions to old installations are a frequent requirement.

The electricity board reserves the right to be notified of any proposed addition to an existing installation, for a number of reasons. First, they must be satisfied that their service cable and meter will not become overloaded. Secondly, they must ensure that any additional connections do not disturb the service to other consumers. (For example, a heavy load, suddenly imposed or removed on a single-phase supply at the end of a long main cable, might cause severe flicker on the lighting systems of all the nearby consumers. Finally, they must test the extension, to ensure that it is safe in every way.)

Extending or repairing old installations may offer many more traps than new work, and the first thing to check is the condition of the existing wiring.

In a later section on testing we shall discuss this matter more fully, but it may be mentioned here that the first test is to see that the consumer unit is not already overloaded, according to the principles set out in an earlier chapter about the loading of the various circuits; and then the insulation of the existing wiring and its earthing condition must be checked carefully.

Too often, in old houses, the wiring (perhaps 30 or 40 years old, and carried out in rubber-insulated wire) has severely deteriorated. The rubber insulation has become brittle and broken in places, and where overheating has occurred (especially at such places as ceiling roses where larger and larger lamps have continually been employed, over the years), the insulation may be in a very poor condition. This may not always be fully revealed with the 'Megger' test for insulation

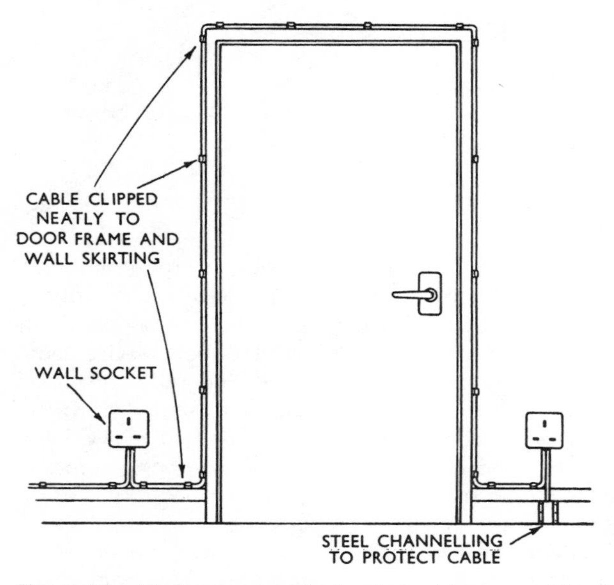

Figure 6.1. Neat arrangement of surface wiring around the doorway and along skirting board, with protection where the wire enters the floor. Bends must conform to *Table* 6.1 (see IEE *Table* 52C)

(see chapter on testing), but such a test should be accompanied by a visual inspection of the wiring in a typical fitting. In such cases, no extensions should be added to such wiring.

Extensions to existing wiring, where they may be safely carried out, may conveniently be made in sheathed wiring, often run on the surface to avoid disturbance to decorations.

This surface wiring is a legitimate method, providing great care is taken to see that the cables are either protected from damage by means of a suitable metal cover, or else run in such a way that they can be clearly seen. A very neat job, with PVC sheathed cable, can be carried out by running the cable along the top of a skirting board, and round a door frame, as shown in *Figure* 6.1. If the cable is subsequently painted to match the woodwork it will scarcely be noticed by the casual eye, yet it is sufficiently obvious to avoid the danger that someone will knock nails right through it.

A point where special care is necessary is where sheathed cable enters a floor. Here it is especially vulnerable. Half round or oval metal channelling should always be fitted at this point, to prevent accidental damage.

Surface-type fittings are usually employed in such work. These fittings have knockouts all round the box, and care must be taken (1) to knock out only the minimum area needed to accommodate the cable, (2) to see that all unused knockouts are left in position, to prevent the intrusion of dust, and (3) that the wires, as made off, are not pinched when the box is screwed to the wall.

When in doubt about the state of an old installation (and tracing wiring runs may be very difficult in old buildings) it may well be better to run the new circuits back to the existing or new consumer units. New units may require to be connected by the electricity board.

Conversion to ring circuit

It is often possible to convert an old wiring system which is still in good order into a ring system, but of course this can only be done if the size of conductor originally used is sufficient to satisfy the Regulations: i.e. it must not be less than 2.5 mm^2.

For example, suppose a house has a 2.5 mm^2 (or the original 7/.029) feed straight from the consumer unit to a 15 A socket-outlet in the sitting room, and another similar feed to another 15 A socket in the dining room (*Figure* 6.2). If these two feeds can be clearly traced at the consumer unit, they

Table 6.1 Minimum internal radii of bends in cables for fixed wiring

Insulation	*Finish*	*Overall diameter for flat twin cable across the widest part*	*Factor to be applied to overall diameter of cable to determine minimum internal radius of bend*
		not exceeding 10 mm	3
PVC or rubber	non-armoured	exceeding 10 mm but not exceeding 25 mm	4
		exceeding 25 mm	6

Abridged from the *Regulations for Electrical Installations, Table* 52C.

Example: If a 1.5 mm^2 twin sheathed PVC cable with a major diameter of 8 mm were used then:

minimum *internal* radius bend = $8 \times 3 = 24$ mm.

can be taken as the two ends of a ring, and brought into one 30 A fuse at the consumer unit.

At the remote ends, surface wiring can be taken from the socket-outlet in one room, all round that room, to feed as many 13 A socket-outlets as desired, and then through the dividing wall into the other room, to feed further socket-outlets in that room, terminating at the second original 15 A socket-outlet position.

A problem often facing those called on to extend old installations is the need to provide an earth connection, when the original installation employed only two-pin socket-outlets, with no earth. In this case, there is no alternative, when installing the 13 A socket-outlets that must be used, to running an earth connection (protective conductor) right back to the earthing point at the consumer's terminal. The size of this conductor must be as laid down in the Regulations, but it must not be smaller than 1.5 mm^2. The temptation to provide a local earth, near to the socket-outlet concerned, by making a connection to any nearby water pipe must be resisted, since the installation must never have more than one earthing point. If protective conductors are added they must be insulated conductors, coloured green and yellow, and be properly clipped and protected throughout.

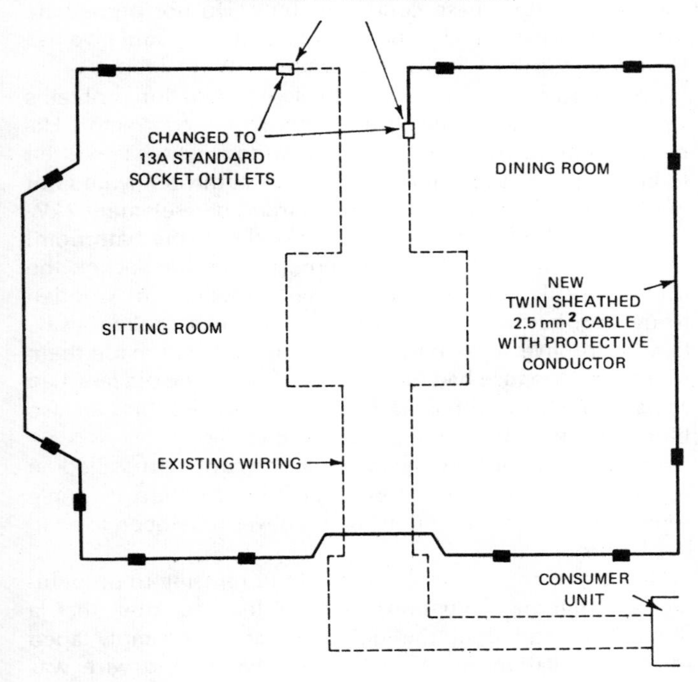

Figure 6.2. Conversion of old wiring to ring circuit by using the existing feeds to socket-outlets.
Note: the existing wiring must be at least 2.5 mm^2 in size (or the old equivalent of 7/.029), and it is assumed that the original socket-outlets are correctly earthed with conductor of 1.5 mm^2 minimum

The tests that must be made before commencing any work on an old installation (see chapter on testing) may show that the wiring is faulty.

In rectifying a fault on the wiring it may be permissible to restore the faulty wiring to its original condition, with good workmanship and good materials.

If non-standard wiring connections are found in an installation (e.g. fuseboards using fuses for phase *and* neutral feeds), then this must be changed to the correct conditions, i.e. fuses in the phase conductor only. Do not perpetuate incorrect wiring, which is basically unsafe and could give rise to serious troubles, including electrocution incidents.

An example of this aspect of obsolete installation problems was recently mentioned to the author by a contractor. His men went to the house of an elderly widow, who had written to him to say that the electric fire in her bathroom would not work. They found an antique, unguarded bare-element 2 kW fire, plugged into a two-pin socket-outlet in the bathroom, obviously with no earth connection. Behind the socket, the rubber insulation on the wiring, installed in wooden troughing, had become so brittle that it had broken away, leaving the live wires bare. Some vibration had made them touch, and the fuse had blown. A nephew of the old lady had replaced the fuse with a much thicker wire, and this had also blown, burning his hand, so he left it alone.

The lady was most indignant when the contractor called on her and said that it would be impossible to effect a simple repair, and the socket-outlet should never have been fixed in the bathroom.

He was, of course, absolutely right in refusing to perpetuate a dangerous situation: but he learned later that a 'handyman' had 'done the job' for £5, and had simply taped up the insulation for a few inches, where the wire was accessible behind the old two-pin socket-outlet, and replaced the fuse. If the old lady had been electrocuted, he would have had much to answer for.

The contractor advised the electricity board that an unsafe installation existed, and they managed to convince the lady that rewiring was necessary.

Warning points to users of electrical installations
The users of an installation should be warned especially about the following points.

It is very unwise to use adapters if it is at all possible to avoid them. They tend to give rise to radio and television interference troubles through loose contacts, and they may well become overloaded, and in consequence overheated,

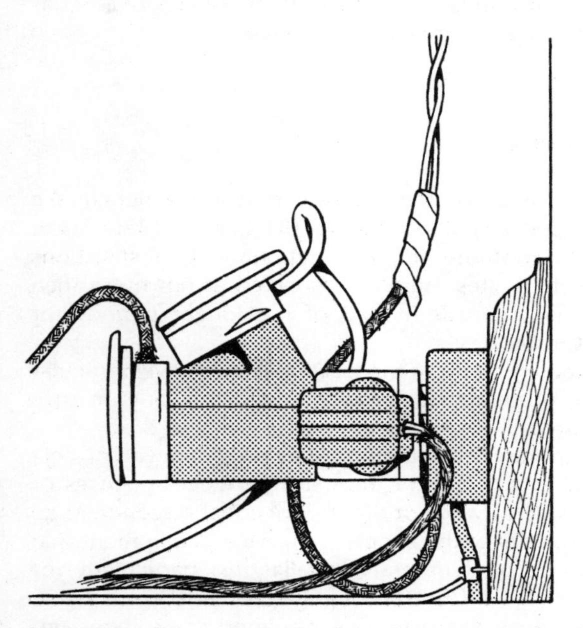

Figure 6.3. This arrangement is to be avoided. This illustrates a mass of adapters, which generally give rise to overheating, interference on radio and television, and the possibility of danger through tripping over flexible wires

and possibly dangerous (see *Figure* 6.3). They give rise to long lengths of flex, which may cause people to trip.

Never extend a flexible lead by simply joining extra flex on to the end and making a taped joint. This can easily pull out and give rise to live wires lying on the carpet for children to touch; and the joint, simply being twisted, is never satisfactory and may itself overheat.

Avoid loose flex wherever possible since it is liable to trip people up or to break or fray when stepped on, and it should never be run under carpets or linoleum, since any damage will not be seen.

Never run an extension by stapling flex on to a skirting board. Flex is not suitable for this type of installation as it has no protection against mechanical damage.

Temporary wiring

The requirements for temporary wiring are generally the same as for permanent installations (IEE Reg. 11–1 refers) but Appendix 16 (Footnote) requires that temporary installations on construction sites must be inspected at not more than three-monthly intervals in lieu of the longer periods for permanent installations.

Certain requirements in the IEE Regulations have particular significance on temporary installations, particularly on construction sites.

Every temporary installation must be provided with protection against excess current (normally by means of fuses or MCBs) and with a switch or other device that disconnects all phase conductors of the supply (*Figure* 6.4). Equipment that has been used on temporary installations, particularly for outdoor use, should be thoroughly checked before being used again; corrosion may have rendered some items unsafe.

All temporary installations must be properly tested and must have the correct values of insulation resistance, earth continuity, and correct polarity (that is, the switches must be in the phase wire and not in the neutral, and socket-outlets must be connected the right way round) exactly as these requirements apply to permanent installations.

For the purposes for which most temporary installations are required, PVC sheathed and insulated cable is usually

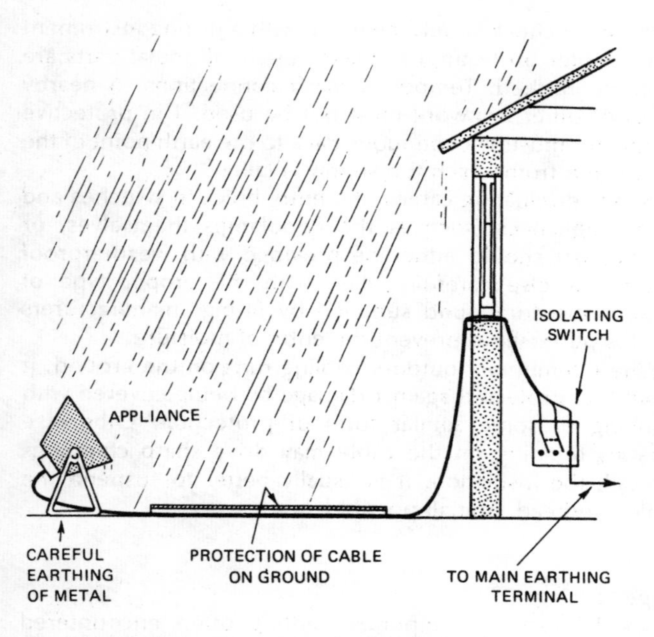

Figure 6.4. Temporary wiring for floodlighting

suitable, but care must be taken to ensure that temporary wiring is not moved by other people so that, for example, wires lie in contact with a hot-steam pipe, or encounter similar hazards.

Temporary outdoor wiring
Outdoor wiring for such purposes as floodlighting must be carried out with special care, since driving rain may well penetrate any of the fittings, and give rise to danger of shock or short circuits and blown fuses.

Extreme care in earthing every metal part of every flood-light or switch is essential. Nothing should be taken for granted – a check should be made with a testing instrument (see chapter on testing) to make certain all metal parts are properly earthed. Temporary earth connections to nearby pipes or other ironwork must not be used. The protective conductor must be continuous back to the earth point of the installation from which the supply is taken.

When terminating cables, the entry holes in switches and other appliances (such as the floodlamps themselves, or fuseboards) should either be provided with weatherproof glands, or else carefully filled with the proper type of semi-plastic compound supplied by fittings manufacturers for this purpose, to prevent the entry of moisture.

Where temporary outdoor cabling runs on the ground, it should be protected against damage by being covered with planking or some similar form of protection. Otherwise persons treading on the cable may drive sharp chippings through the insulation. It is usually better to suspend the cable overhead, if at all possible.

Stage lighting

A special case of temporary lighting often encountered relates to wiring for theatrical purposes in public halls.

If the supply is to be taken from 13 A socket-outlets (one of which is adequate for 3 kW of lighting – thirty 100 W lamps), care must be taken that suitable isolating switches are provided for the circuits connected to each separate socket-outlet.

Dimmers

Dimming of stage lighting is often required. Dimmers using variable resistances are available, and must be of the rating needed for each circuit, otherwise they will be overloaded and could become dangerously hot. The easiest method of providing for dimmers is to install an additional two-pin socket-outlet in series with the phase wire and arrange for it to be controlled by a switch. The circuit is shown in *Figure* 6.5.

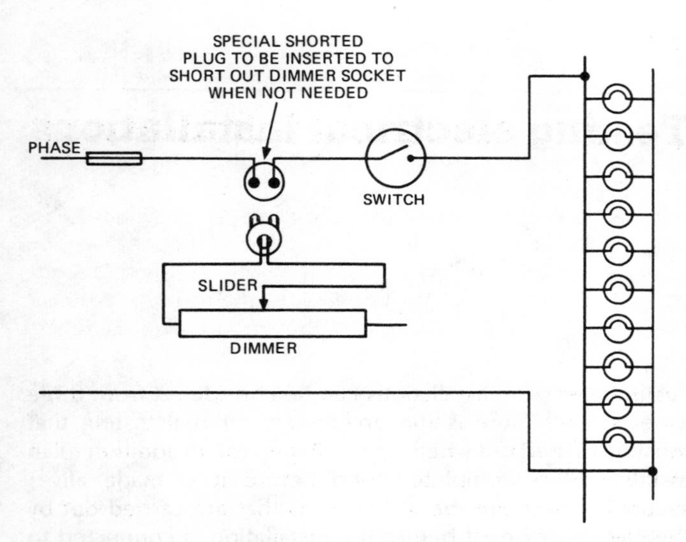

Figure 6.5. The provision of an additional socket, normally shorted out by a shorted plug top, so that a dimmer can be inserted when required, as in a village hall wiring system where stage lighting is occasionally needed

Floodlight and spotlight connections

When supplying floodlights and powerful spotlights, the proper method is to terminate the main wiring 2 m or more away from the floodlamp and then to feed the lamp, via the socket-outlet, with special heat-resistant cable.

7

Testing electrical installations

Testing electrical installations can be considered from three aspects: first there is the preliminary internal testing that must be carried out when an installation (or an addition to an installation) is completed, and before it is made alive; secondly, there are the official tests that are carried out by the electricity board before the installation is connected to the mains; and finally there is the testing necessary to find faults, or to ascertain the condition of an old installation.

Equipment needed

The basic instruments needed for testing and checking are illustrated in *Figure* 7.1. They comprise:

(1) A universal test meter.
(2) A bell and battery set for tracing wires.
(3) A 'Megger', combined with a protective conductor continuity tester.

The universal test meter

The universal test meter is a device whereby a single scale can read volts, amperes and ohms, by adjusting the knob with appropriate markings. An instrument of this kind can be invaluable for locating faults. It must be capable of reading up to 500 V a.c. or d.c., and the resistance range should be capable of reading down to less than one ohm.

The resistance range on such meters is operated by means of a small dry battery, inserted into the instrument in a similar fashion to that employed in portable transistorised radio receivers. To allow for variations in battery voltage, an adjusting knob is provided. When the two leads from the instrument are clipped together, obviously the resistance between them is practically zero, and the adjusting knob is

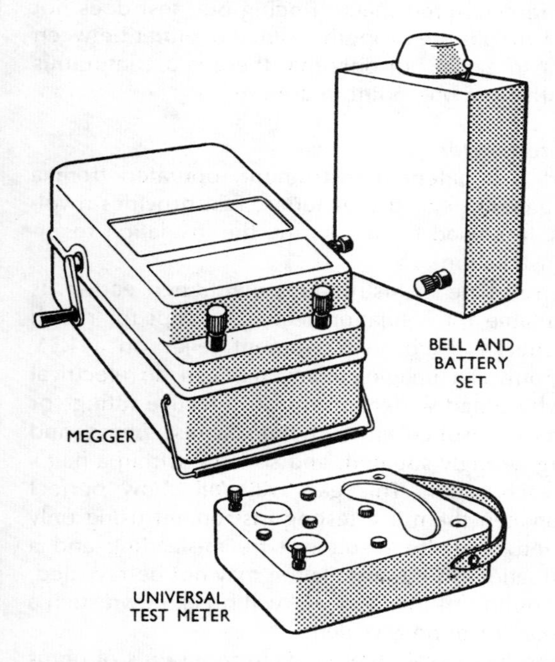

Figure 7.1. The basic instruments needed for testing and checking: the universal test meter, the bell and battery set, and the 'Megger'

moved until the needle reads zero. This adjustment should be carried out before every set of resistance tests is undertaken.

The bell and battery set

For rapid checking of circuit connections, a bell and battery set is extremely useful. Two torch batteries are connected to a bell, and the assembly is connected to two probes or clip leads, so that when the leads are joined together the circuit will be completed and the bell will ring. A long lead will enable circuits in all parts of a domestic installation to be checked out rapidly, or 'rung out'.

It must be remembered that a 'ringing out' test does not prove that the circuits are properly insulated, either between conductors or to earth, but only that there is a continuous conductor path from one point to another.

The 'Megger' test meter

The 'Megger' is a patented instrument, operated from a hand-driven generator or dry batteries, that provides a voltage of 500 V. It is used for measuring the insulation resistance of the installation.

The 'universal' type of instrument, mentioned earlier, is entirely unsuitable for insulation testing, except for rough preliminary checks. Such an instrument relies on a 1.5 V battery, and one can imagine many parts of an electrical installation where bad workmanship or defective fittings or appliances have resulted in two bare wires, phase and neutral, being wrongly situated, and so lying within a hair's breadth of each other. This gap will still show perfect insulation between them if a testing instrument using only 1.5 V is applied, and the result will be misleading and a dangerous situation on the installation may not be revealed. But the 500 V output of the 'Megger' will break down such a gap, and the fault will be revealed.

The 'Megger' has a scale that reads in thousands of ohms or in megohms (millions of ohms), and it is so arranged that once the turning of the handle has reached a certain speed, no increase in speed will affect the reading.

Some types of 'Megger' have a second instrument incorporated in the same case. This instrument measures the impedance of the earth loop at any point in the installation.

With alternating current, as mentioned earlier, there are factors other than pure resistance that affect the amount of current flowing through a circuit. Where coils are concerned, the electromagnetic effect may mean that there is greater opposition to the passage of the current than that due to resistance. The resulting combined opposition to the passage of current is called impedance, a term applicable to a.c. circuits only.

The impedance of the earth loop – that is, of the path carrying earth leakage current from any point on the installation to the supply transformer – includes the resistance of the protective conductors in the installation, the resistance of the earth electrode or other earth connection to the general mass of earth, and the electromagnetic effects (which may arise in any part of this circuit) mentioned earlier.

The maximum permissible values of earth-loop impedance are given in IEE Regs 413–3 and 4 and Appendix 7. To check that it is below this impedance, a suitable instrument (which may be incorporated with the insulation-testing 'Megger') must be used.

Preliminary testing

These tests are *not* intended to be the 'official tests' of the installation, and are only intended as a guide to the condition of the wiring.

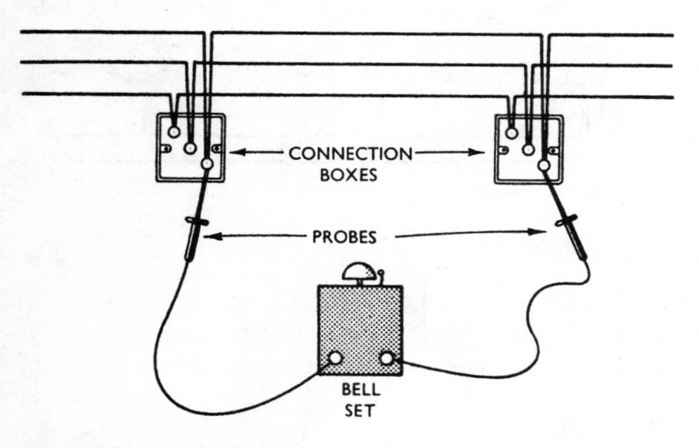

Figure 7.2. Continuity test using bell set

All preliminary testing is carried out before the main fuses have been inserted.

Continuity test

The first test is a simple continuity test to ensure that all connections have been properly made. The bell set (*Figure* 7.2) can be used for this, and also to check the proper connection of the switches, thermostats, time switches and other devices, to ensure that they are all on the phase side of the circuit. This can be tested by checking continuity from the neutral connection of the circuit being tested, at the consumer unit right through the thermostat, switch, or other interrupting device, making sure that the switch breaks the phase wire and not the neutral wire.

Polarity test

Next there is the question of polarity – socket-outlets having their connections made the right way round. On the ordinary 13 A fused plug socket-outlet, looking at the face of the socket-outlet, the phase connection must be on the right, the neutral on the left and the earth at the top. A simple continuity check from each socket back to the phase fuse will ensure that this is correct (see *Figure* 7.3).

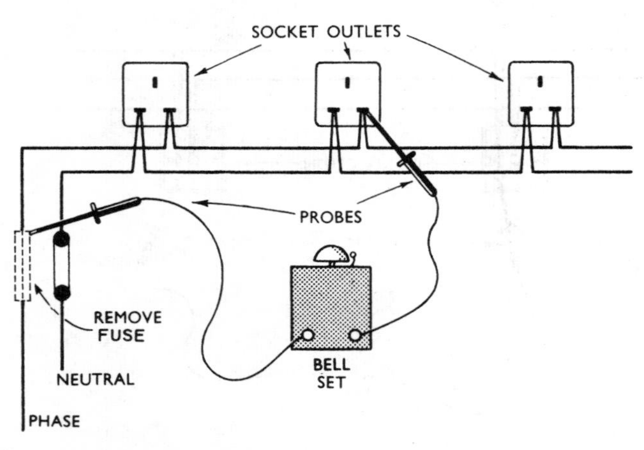

Figure 7.3. Polarity test with bell set

Protective conductor test

The next step is to check protective conductor continuity. For this purpose, an *approximate* check that no gross errors have occurred can be made by means of the universal testing instrument, arranged on its ohm-reading scale.

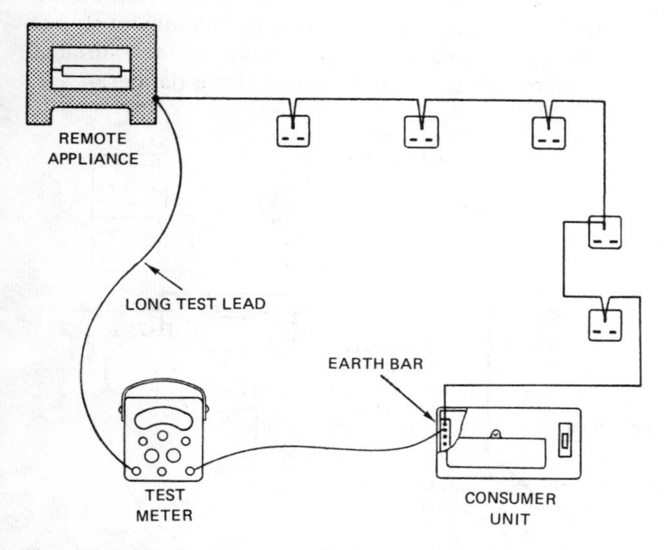

Figure 7.4. Preliminary test of protective conductor resistance

It will be recalled from the section on earthing, that all exposed metal that can possibly become connected to a live circuit must be effectively earthed, in such a way that the resistance of the protective conductor is of low enough value, to carry fault currents without danger.

With one lead of the testing instrument on the earth point, and the other arranged as a probe that can be attached to all parts of the fixed appliances, or the portable equipment plugged into the socket-outlets, the protective conductor resistance can be checked with the universal instrument (*Figure* 7.4).

However, this cannot be taken as the full, official test. The reason for this again hinges on the fact that the universal instrument only has a 1.5V battery. Suppose a connection was badly made, so that the wires were simply touching each other, very lightly. This would mean that a 1.5V battery would show a good connection, but if a heavy current should pass through this circuit, the connection would obviously not be good enough: arcing and ultimate melting of the surfaces in contact would occur, with consequent fire danger.

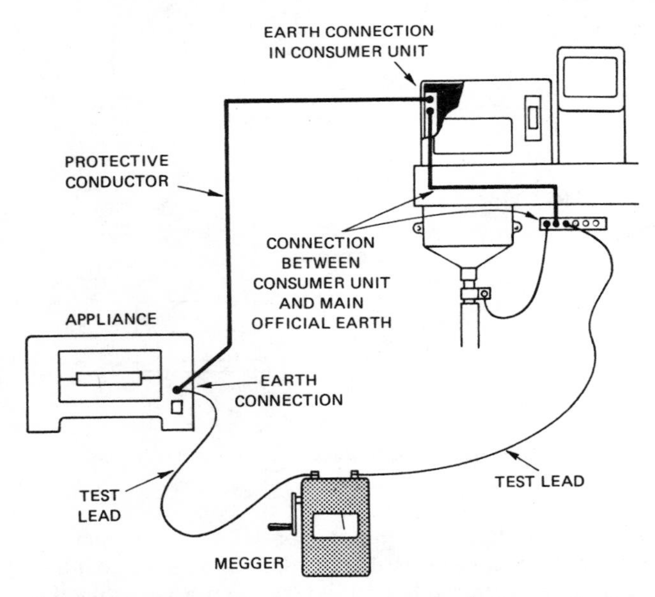

Figure 7.5. 'Megger' test on protective conductor continuity

In fact, the Regulations specify that when testing the protective conductor, alternating current of a magnitude approaching one and a half times the rating of the circuit under test shall be used, with a maximum of 25A. As mentioned above, only a properly designed instrument can provide a test current of an appropriate value to make sure that the slightest defects in the protective conductor system are observed and can be detected (see *Figure* 7.5).

Insulation test

This test ensures that the insulation throughout the whole installation has not been damaged in any way, and for low voltage circuits, that is up to 1000 V a.c., the test has to be carried out, according to the Regulations, with a direct current voltage not less than twice that which will be normally applied to the installation, although it need not exceed 500 V, for installations rated up to 500 V.

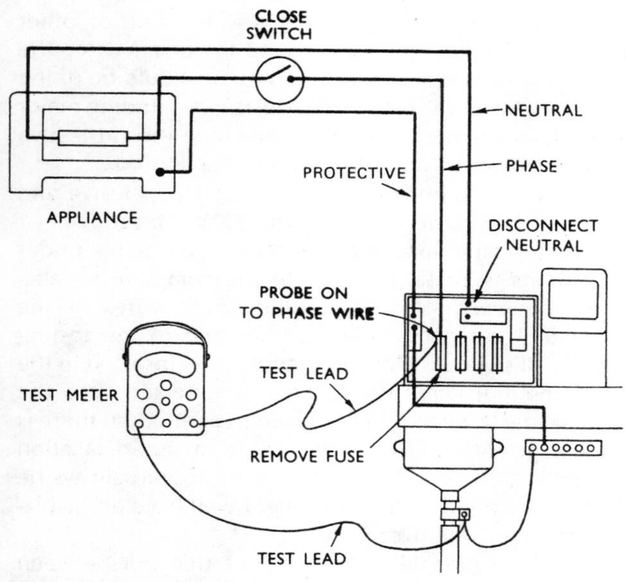

Figure 7.6. Insulation to earth test circuit

The figure of 500 V is that which is usually used, with the aid of the 'Megger'.

Again, the universal type of instrument would not be suitable, even using its resistance scale. If, for example, the wire running through a conduit had been strained over a sharp edge incorrectly left at the end of a tube, the bare conductor might be within a tenth of a millimetre of the earthed metal, and yet the 1.5 V output of the instrument would not break down this gap and reveal the defect.

The insulation resistance is measured on each circuit by closing all switches on all appliances in circuit, and with the neutral connection disconnected (*Figure* 7.6). Under these conditions the circuit will be complete throughout the phase wire, the switch, the appliance, and the neutral wire back to the consumer unit.

The universal tester again may be used to provide a rough check. One lead should be clipped to the official earth point (or for a rough preliminary test) to a water pipe or other convenient earth, and the other lead to the phase wire. The reading on the ohm scale of the instrument should be of the order of 1 or 2 MΩ. This test is only suitable for finding major errors, such as the live wire firmly touching the protective conductor connection in a fitting or connection box.

But it must be emphasised that the only conclusive and official test is that carried out with the 500 V 'Megger'.

The bare minimum insulation resistance acceptable under the regulations is 1 MΩ, applying to the complete installation. This means that when all the phase wires at the consumer unit are connected together and to the testing instrument, all switches closed, all appliances inserted in the circuit, all neutral wires being left disconnected, and the other end of the 'Megger' is connected to earth, then there is a minimum of 1 MΩ between the whole of the installation taken together, and earth. A higher value should always be aimed at, and a good installation might well have an insulation resistance of well over 5 MΩ.

A second insulation test should be carried out between conductors (see *Figure 7.7*); for the purpose of this test all lamps and appliances are removed, or isolated by opening their local switch, and the test instruments' leads connected to the phase conductor and the neutral conductor, the minimum reading being 1 MΩ.

If for the purpose of either of the above tests, equipment is removed, then it should be tested in a similar manner

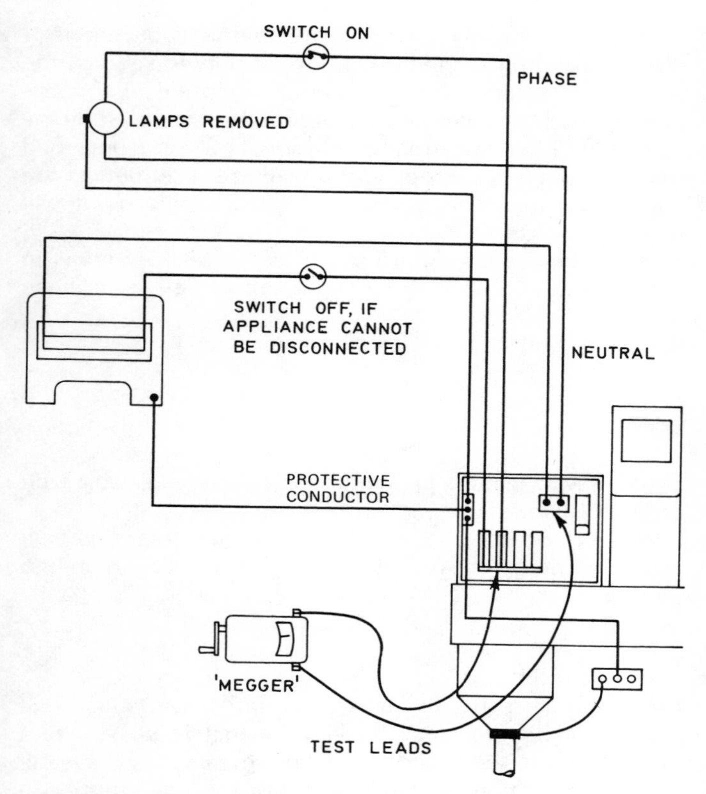

Figure 7.7. Insulation test between conductors

separately, and in this case a minimum reading of ½MΩ is permissible.

Official testing

Before connecting the supply, the electricity board will inspect and test the installation and check that the wiring is satisfactory.

They are not concerned with the internal connections of the installation itself and it is not their duty to see that the proper lamps light when the switch is closed, or that appliances such as water heaters and the like, installed by the electrician, operate correctly. They are, however, concerned with the safety aspect of ensuring that socket-outlets are connected up the right way round, and this point will therefore be checked.

They will also check that there are no socket-outlets in bathrooms, and that other safety measures have been observed.

Tracing faults

Suppose now that the preliminary tests have revealed a fault on the system.

With the aid of a bell set, and a universal meter, the fault can be traced quite easily by splitting the installation up into sections.

Insulation fault

Let us suppose that the fault is an insulation fault – that means that the whole installation shows that the phase wire is in effect connected to earth, and therefore does not have the required insulation resistance of at least 1 MΩ with respect to earth.

The circuits comprising the installation can easily be separated out at the consumer unit. Taking one circuit at a time (the fuses, of course, still remaining withdrawn), disconnect the neutral wires and thus both ends of the circuit are clear of any other connection. Then, with all the switches closed and appliances connected, check each circuit under these conditions, until one (or more) is found on which there is a fault (see *Figure* 7.8).

Concentrating on this circuit (all others being healthy) first the appliances may be disconnected, one by one, until a possible faulty appliance shows up. If no appliance is faulty then proceed as follows:

Assume for the moment that there is only one fault in the circuit. Go to the first disconnection point on this circuit (which might be a joint box on the sheathed cable run), and disconnect both wires.

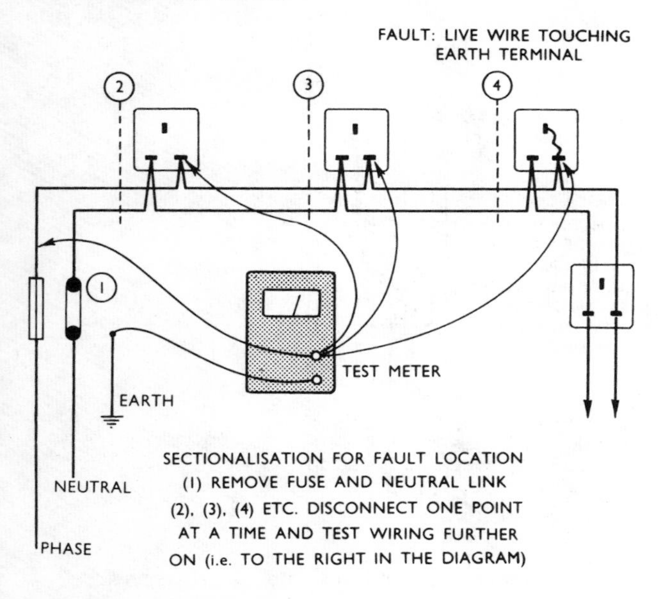

Figure 7.8. Principle of sectionalisation when testing for faults

Return to the consumer unit and once again check both the phase and neutral wires for insulation to earth. Suppose that they are both 'good', the fault has thus been removed by this disconnection. It is obviously beyond the point of disconnection.

Transfer the testing instrument to the first joint box, which has already been opened, and proceed further to the next joint box or switch, the next break point on the circuit, and disconnect the wires in this second joint box.

Test again. If the fault is cleared, proceed further down the circuit. If it is not cleared, then obviously it lies in the section between the testing point (at this first disconnection point) and the further point at which the circuit has been broken.

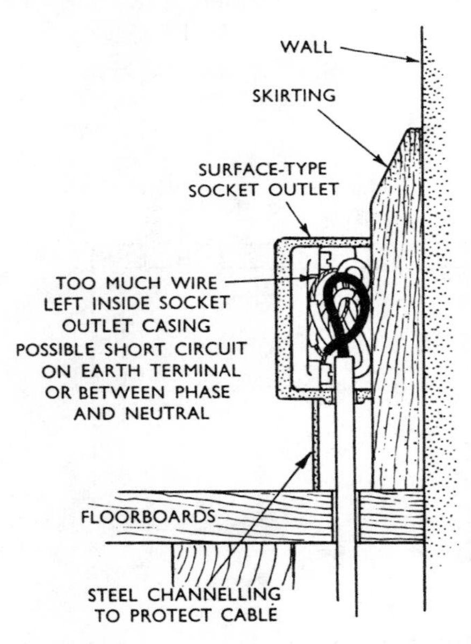

Figure 7.9. Common causes of faults on electrical systems

Suppose the fault *is* cleared, then proceed further and go beyond the switch (or second connection point) which may perhaps bring you to some fittings or socket-outlets. These must be methodically disconnected, one by one, testing after each disconnection, until the faulty unit is isolated. It must then be inspected to find where the fault lies.

One of the most common causes of faults in domestic installations is that the wiring behind a socket-outlet or a switch has been incorrectly made off, so that either too much bare conductor exists, and is touching the earthed metal or the neutral wire, or, on the other hand, the switch or other fitting may be incorrectly assembled, possibly so that two wires are incorrectly inserted into one terminal (see *Figure* 7.9).

Suppose the fault has been found to be in a run of cable. If the installation is carried out in conduit, it will not be very difficult to pull out the cable in that section and inspect it, and the fault will probably soon be found. It may well be due to the insulation being damaged as the cable was drawn into the conduit.

If the installation is carried out in sheathed wiring embedded in plaster, then it will be necessary to break out the section containing the fault and replace with with new wire. It is unwise to attempt to locate the fault and to patch it up. New cable must be used.

Up to now it has been assumed that the fault is one of bad insulation. There are, however, two other kinds of faults.

Short circuit

First, there is a short circuit (*Figure* 7.10). Both the neutral and the phase wires may remain well insulated from earth, but are short-circuited to each other.

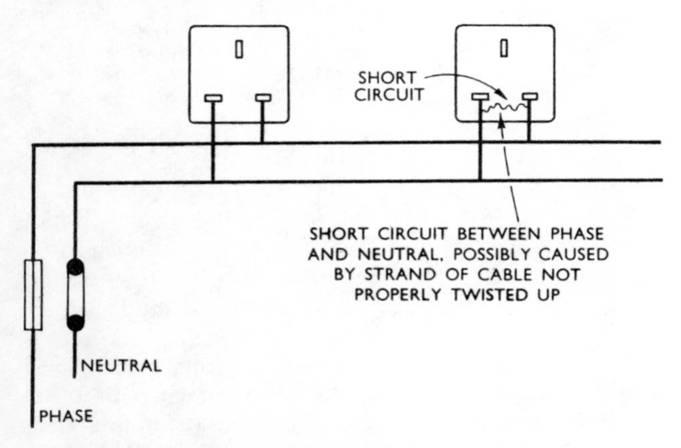

Figure 7.10. Testing for a short circuit

In the majority of cases this is due to incorrect connections in fuses, switches or fittings. It is unlikely that the wires within a conduit, or inside the sheath on a sheathed cable, have come into contact with each other without at the same time going down to earth, although this type of fault could not necessarily always be ruled out. A short circuit could occur if severe mechanical damage has occurred to sheathed cable, perhaps under a floor, where for example, a workman from another trade, such as a plumber fitting water pipes, or a gas fitter, has inadvertently severely manhandled the cable.

The short circuit can be isolated by the same methodical sectionalising test methods as those outlined above.

Open circuit

Another type of fault is the open circuit (*Figure* 7.11). Here again, in this case where there is no circuit between two

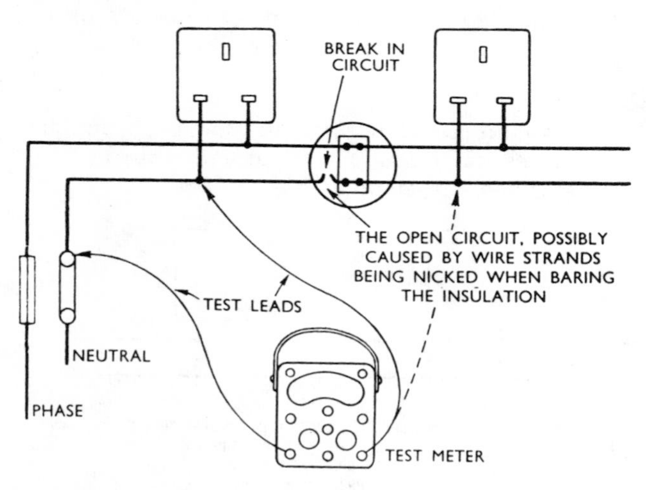

Figure 7.11. Testing for an open circuit

points, the fault is most likely to lie in a switch or other fitting. There have been cases where open circuits have arisen when fittings have been incorrectly made off, and the electrician, roughly baring the ends of the wire, has nicked through all the strands, and then when the tension comes on the 'tail' which has been inserted into a switch or socket-outlet or some other fitting as the unit is screwed together, the weakened wires break away, leaving an open circuit.

Cables that have been badly kinked before installation could give rise to open circuits, as the kinking may have broken the copper conductors.

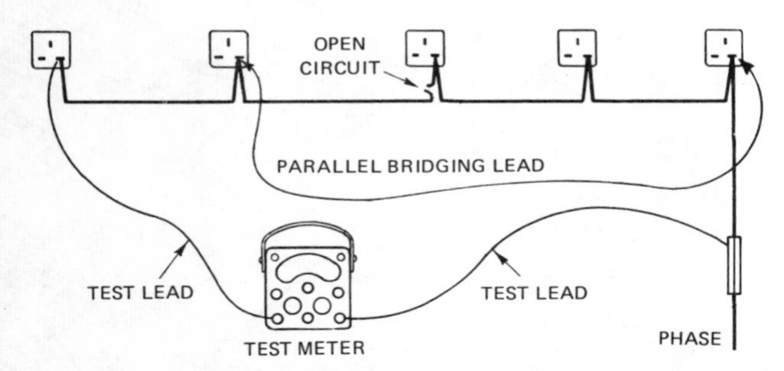

Figure 7.12. Locating an open circuit by paralleling sections of the wiring with a long test lead

The open circuit may not be so easy to find, under certain circumstances, as the other types of fault, but with the aid of a long test lead, it is not difficult to parallel each installed wire by an external lead across the gap between, say, two junction boxes, or the consumer unit and a fitting, and if a complete circuit is obtained in this way, shorten the test lead until the section or fitting in which there is an open circuit (see *Figure* 7.12) is found.

Typical faults in installations

Incorrect system of fusing

Where the normal supplies of phase and neutral are concerned, double-pole fusing has frequently been found. This is most often found in a final circuit where there is a circuit distribution board.

Insufficient protection for sheathed wiring
It is frequently found that when sheathed wiring is out of sight, it is also out of the electrician's mind, in the sense that he has not provided any protection at all. It is just as

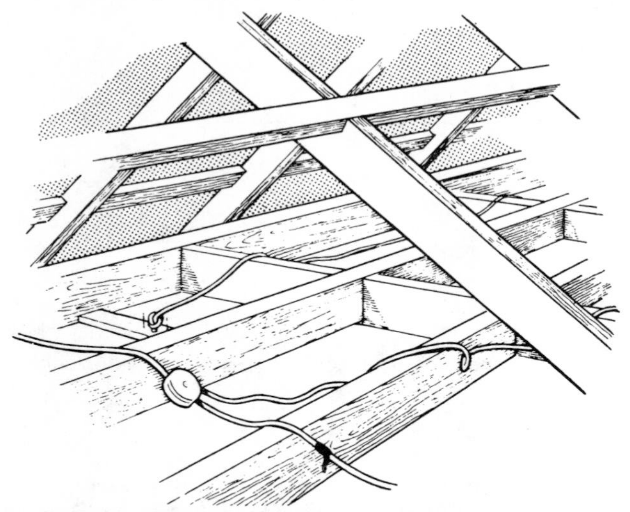

DEPRECATED PRACTICE OF UNSUPPORTED SHEATHED WIRING IN LOFT, WITH JUNCTION BOX UNFIXED

Figure 7.13. Bad wiring in lofts and cellars is sometimes found, with the wiring unsupported, the junction boxes unsecured, and the cable unprotected

important that sheathed wiring should be protected from mechanical damage in such places as lofts and cellars, which are infrequently used, as it is in cases where the wiring is obviously visible (see *Figure* 7.13).

Conduit installation troubles
Insufficient attention is often given to the filing away of sharp edges on the ends of conduit piping, and the lack of the

provision of bushes. This means that all cables drawn through this particular piping may well be damaged, and liable to subsequent failure.

Insufficient attention has often been paid to the protection of conduits, ducts, and trunking systems against the entry of water.

Overcrowding of cables in conduits and trunking

No more wires than those laid down in the various Regulations should be accommodated in a given conduit, and if these figures are exceeded there may be trouble due to overheating and possible failure, and in any case the cables will have to be derated (used at a lower current rating than normal).

Incorrect cable for heating appliances

Heat-resistant insulated cable should always be used for connections to immersion heaters, thermal storage heaters and indeed any appliance that gets hot. It has often been found that ordinary PVC insulated cable is employed.

Omission of any identification at the consumer unit

If the circuits are not identified and the type of fuses that are appropriate for each circuit clearly stated, it will be much more difficult to put right any trouble of any kind that may arise on the installation, now or in the future.

Failure to use flameproof or intrinsically safe equipment

In places where flammable liquids are stored, or where, for example, the gas bottles used for welding, or for domestic gas purposes in conection with caravans and the like are located, properly designed flameproof equipment and wiring are necessary.

Insufficient attention to the bonding of all metalwork to earth

Every piece of metal in any way associated with an electrical appliance should be bonded to earth, and it is often found that this is omitted.

Omission of protective shield or skirt to lampholder of bathroom, in lighting pendant or fittings
All lampholders in bathrooms must be insulated and fitted with a protective skirt.

General note on testing old installations

When rectifying a fault on old wiring, do not start by assuming that the supply is the usual single-phase and neutral 240 V a.c. system. Several other systems are still in use in remote parts of the country.

The first test, therefore, should be to find out the system in use. The nameplate on the meter, plus the information given on any lamp that may be in use, will provide a useful starting point. The meter will show if the system is a.c. or d.c., and the frequency employed, and probably the voltage as well, and in any case the lamp will show the voltage.

But the universal test instrument should be applied to check the voltage from each supply wire, separately, to earth.

It should always be assumed that the system is wired up wrongly: never take it for granted that single-pole fusing (as standard on modern installations) is used or that single-pole switches are placed in the phase conductor, or that the neutral wire is in fact at or near earth potential, or that any earth connection at all does in fact exist.

Test records
Keep a record of all tests, with the date and leave a copy with the installation. It will greatly assist anyone working on the installation at a later date.

Testing an installation after fire or water damage
If a fire has occurred in some part of a house, or if, for example, flooding has taken place, or if a tank in a loft has overflowed, a check should be initiated immediately before any current is used.

The main switches on the consumer unit should be opened at once, and the 'Megger' should be applied to each circuit to make two insulation tests. First, when the fuses are removed, access can be obtained to the phase end of each circuit and the neutral connections can be taken out one by one, and with all appliances disconnected but the switches on the circuits closed, an insulation test can be carried out between phase and neutral on each circuit. Then a similar insulation test is carried out between each conductor and earth.

With plastic wiring, heat may have caused considerable damage to the wiring and major renewal operations will have to be commenced. On the other hand, water damage will not have affected the insulation itself, but may have given rise to pockets of water in socket-outlet boxes, switches and other fittings. This means that each of these must be opened and dried carefully, and for this purpose an ordinary hair dryer fed from some circuit in another part of the house which is in good order, may conveniently be used. It is probable that at least one circuit may have escaped major damage.

8

Miscellaneous aspects of installation practice

Outdoor wiring

Apart from underground wiring connections between buildings (which may be carried out by means of mineral-insulated or PVC/SWA/PVC cable, properly protected by means of bricks or concrete slabs), outdoor wiring should employ a catenary, or suspended system (*Figure* 8.1).

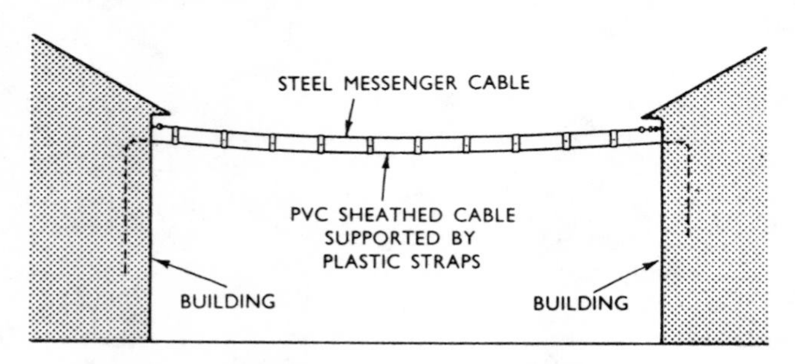

Figure 8.1. Catenary wiring system for outdoor work

There are two methods of catenary suspension. In one method, a special cable is employed, using a chlorosulphonated polyethylene (CSP) sheathed cable with a built-in steel wire for suspension purposes. In the second method, a separate steel wire is employed, and the special outdoor cable is suspended from it by means of straps made from plastic material in ring or tape form.

It is permissible to span a gap of 3 m between adjacent buildings by means of a span of PVC sheathed and insulated cable, carefully clipped at each end to ensure that its own weight, and possible movement in the wind, will not cause the cable to be damaged at the clips.

Bell, television and telephone wiring
Wiring for bells, telephones, radio, television etc. does not fall within the scope of the IEE Regulations for Electrical Installations but the requirements for the segregation of these services are given in Regs 525–1 to 9. In general, wiring for telephones, radio and television should be kept entirely separate from any wiring carrying mains voltage. Bell wiring should also be kept separate from mains wiring unless the bell wiring is insulated for mains voltage.

Bell wiring
Most bells or chimes are supplied from a bell transformer which must be properly fused on the low voltage (240 V) side. The first test should ensure that both windings are continuous and the second test should ensure that the insulation between the 240 V side and the extra-low-voltage side is in good order (see *Figure* 8.2).

Bell-circuit wiring is usually carried out in twin plastic flex, which may be run in plastic conduit embedded in the plaster or neatly tacked with insulated staples on the surface. Joints in such wiring should preferably be soldered and taped to make sure that there is minimum resistance at any jointing point, but ordinary mains-type plastic connectors may of course be used. As stated above, bell wiring may be run in the same conduits as mains wiring, provided the bell wiring is insulated for mains voltage.

Telephone wiring
In new property it is often advantageous to provide concealed conduits, with draw wires, for telephone wiring so that this can be installed at a later date without unsightly surface wiring. These conduits should run from the lead-in position to the anticipated telephone points; they should be

quite separate from any other conduits and must not contain other wiring.

It is permissible for certain approved types of equipment and wiring, for connection to the British Telecom public network, to be supplied and installed by parties other than

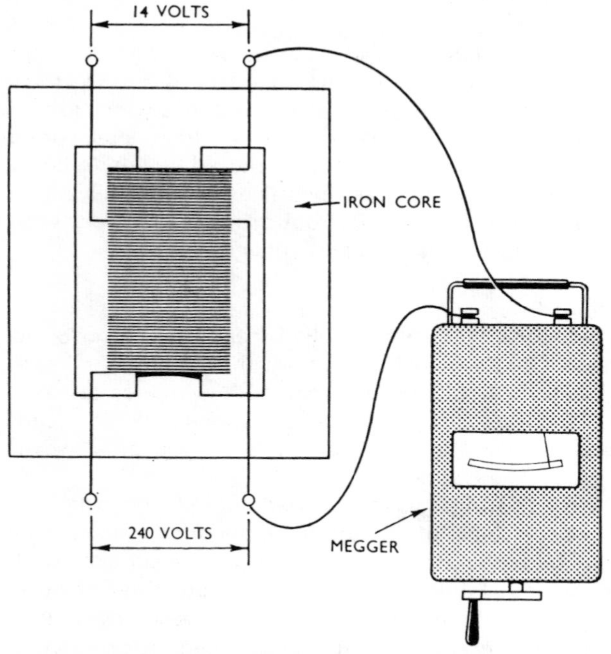

Figure 8.2. The bell transformer, checking insulation between high voltage and low voltage sides

British Telecom. When this is contemplated the appropriate authorities should be consulted and any necessary approvals obtained before the work is commenced.

Television and radio circuits

It is becoming increasingly common for television and VHF aerial down-lead circuits to be installed as part of the wiring of a building, and to terminate at high-frequency sockets

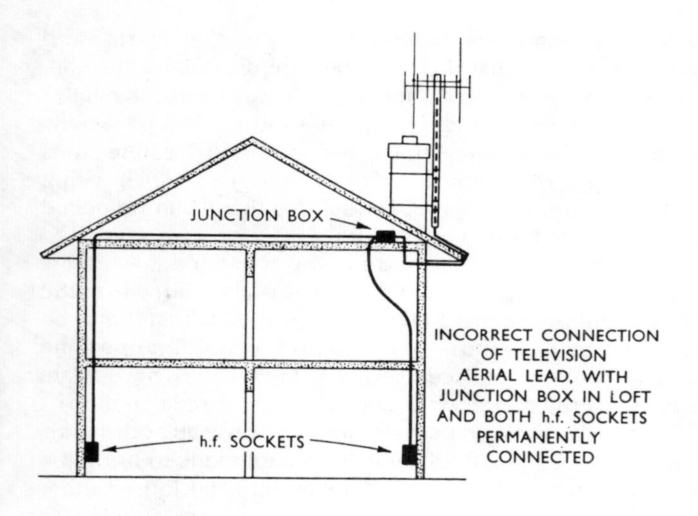

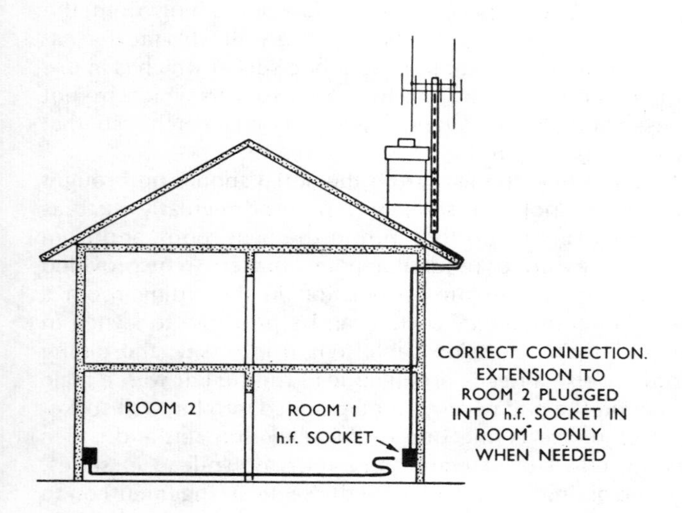

Figure 8.3. Incorrect and correct connections of television aerial leads to high-frequency sockets in various rooms in the house

which are themselves mounted on plates that fit standard wiring boxes, so that they may be sunk flush into the wall, enabling the television set aerial lead to be plugged in neatly.

Low-loss television cable may be used for two purposes: for the television connections, and for the VHF connections for FM radio reception. It is common to install such wiring below the plaster in plastic conduit, so that it can be pulled out if trouble develops.

There is no technical reason why this wiring, which is usually plastic covered, could not be installed directly in the plaster, but if so there is always the possibility that it may be damaged by nails driven into the wall. It will then become very difficult to replace, and any fault could necessitate breaking away plaster walls and subsequent redecoration.

There is a most important point to be watched in this connection. The method sometimes adopted is to bring the lead from the aerial on the chimney into the loft or other similar area, and then to join on to it, leads running to high-frequency sockets, in for example, the dining room, the sitting room, and possibly a bedroom as well. This means that the aerial lead has connected to it one socket which is in use for television reception and two other sockets which are not in use, and this frequency causes a 'mis-match', so that reception is impaired.

To avoid this, the lead from the aerial should be brought down to the point most likely to be used regularly, such as the sitting room, and then run to the next room, and from there to the third or other plug points that are to be provided (see *Figure* 8.3). At the termination in the sitting room a special high-frequency switch can be provided to switch in the additional television cable length into, say, the dining room, or else a neatly arranged lead can be left with a male socket on the end which can be plugged into the wall socket in place of the connection to the television set, and which has the effect of extending the aerial down-lead in series, into the dining room, and (with the same arrangement) on to a bedroom or other receiver point. In this way, the best reception will be obtained without troubles due to mis-matching.

Audio-frequency (sound) circuits

In some installations, sound recording enthusiasts (perhaps in connection with the making of sound films), may like to have facilities whereby one room can be used as a studio and the output from the microphone and amplifier taken to other rooms where recording can be carried out without the noise of the instruments disturbing the studio atmosphere.

This facility can be provided very easily if it is foreseen during the installation work. All that is needed is to use the same type of VHF television cable as used for the television circuit itself, and to provide links between the rooms, terminating in each case on a flush-type high-frequency socket-outlet inset in the wall, so that microphone or loudspeaker circuits can be plugged in as required without running very long leads (which may pick up hum or give rise to attenuation problems) through passages and under doors.

Motor circuits

IEE Reg 552–3 requires that every motor exceeding 0.37 kW should be provided with control equipment incorporating protection against overcurrent in the motor.

Large motors require starters which limit the starting current, but it is now quite common for motors, up to 3.5 kW or above, to be switched direct to the mains.

The types of motor used in such domestic devices as washing machines, vacuum cleaners, refrigerators and power drills do not involve any special problems. However, for motors approaching 750 W and larger (such as those used for lathes in small workshops), there is one factor that needs to be taken into account.

When a motor starts, it may well take ten or twelve times its normal operating current. If the fuse size is based on the running current only, the current rush at starting, although of brief duration, may blow fuses unnecessarily, since the installation will not be harmed, nor the wiring overheated, by this excess current of very brief duration. To give an example, a 750 W single-phase motor at full load will take just over 3 A, but when starting it may take considerably more current than this and the makers recommend that a 10 A fuse should be used.

Motors that have to start very frequently may give rise to difficulties as the excess current during the starting period heats up the fuse more and more, and ultimately it may blow, even though there may be no defect in the installation.

One method of avoiding this is to install a motor starting unit, which is usually available from the manufacturer of the motor, and comprises a pushbutton-operated starting switch. It is usually equipped with a special overload device to prevent damage to the motor or to the installation. These devices often take the form of a thermal overload. A small heating element, suitably proportioned, is connected into the motor circuit, and operates a bimetal strip in much the same way as the thermostat mentioned earlier. Operation of the bimetal strip trips out the switch.

For domestic and small workshop applications up to about 0.50 kW, single-phase motors are the most commonly used. They present no problems in regard to their connection to the supply.

For larger drives, the 3-phase motor requires a special 3-phase supply, and a 3-phase isolation switch must be provided, even if the starting arrangements are incorporated within the machine to be driven.

Transformers

The transformer is a most useful static device, with no moving parts and therefore no need for maintenance, used to change one voltage to another. Basically, it has remained unchanged since it was first developed by Faraday in 1831.

The transformer consists of iron stampings made from thin plate, arranged in the form of a ring (*Figure* 8.4). On one side is a winding having, let us say, 2000 turns, and on the other side a winding having, say, 1000 turns.

If the 2000-turn winding (the primary winding) is now connected to a supply of alternating current, then a voltage will appear on the 1000-turn winding (the secondary winding) which is exactly one-half of that applied to other side. The voltage can be stepped up or down at will by changing the 'turns ratio', providing the coils are suitable for an application at the voltage desired.

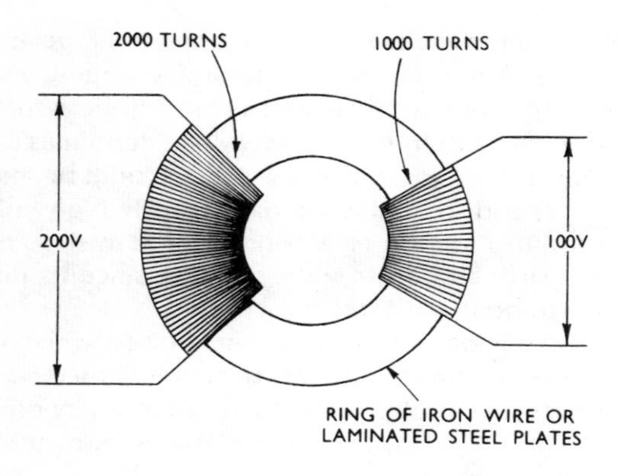

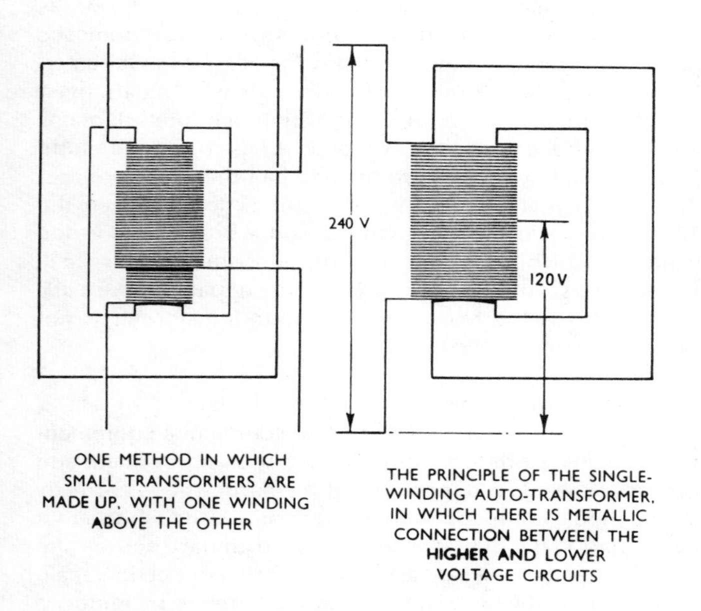

Figure 8.4. The principle of the transformer: the two-winding transformer and the auto-transformer, which is not permitted for normal use in domestic premises

The most common example of transformer in domestic use is the bell transformer, giving an extra-low voltage from the 240V supply. In this case, it would obviously be wrong to connect 240V mains to the extra-low-voltage terminals as the insulation of the extra-low-voltage winding would be insufficient for 240V, and in addition a dangerously high voltage would be produced on the 240V winding, and this too could flash over to earth and destroy the winding since its insulation would also be insufficient.

There are regulations concerning transformers which insist that the insulation between the two windings should be of at least as high a value as the insulation of the primary or 240V winding to earth. This prevents the mains voltage from penetrating to the low-voltage side.

There are, however, certain types of transformer known as auto-transformers, which must not be used for domestic purposes, although they have many applications in industry. In this type of transformer there is a single winding arranged on one side, as it were, of the iron ring, and the full mains voltage is applied across the winding. A tapping is taken from it, at, say, half-way down, to provide half the mains voltage. This means that there is physical connection between the 120V circuit and the 240V circuit, and a breakdown in the winding (which does happen from time to time on small transformers) could result in 240V being applied to the 120V circuit. This is the reason why the auto-transformer is not permitted in domestic premises.

Continental equipment

The marking of the wires connecting continental equipment to the mains is often different from that used in Britain, and care must be taken to ensure that the earth wire is correctly ascertained. Some continental machinery uses white earth wires, and many domestic appliances manufactured on the continent and in the Far East have no earth connection at all. In such cases, the two-core flex should therefore be removed and a three-core flex substituted, but there a further problem arises. Within the device itself – say, for example, a food mixer – it may not always be possible at first glance to see where to attach an earth connection in such a way that it ensures that all metal parts that can be touched are connected properly to the earth wire. This means some care and thought and perhaps the need to arrange some bonding wires within the appliance to make certain that all metal parts of all kinds within the appliance are in fact joined together and to earth. However, the appliance may be of the double-insulated type (see page 9), and this is then unnecessary.

Cupboard door switches

To provide an arrangement whereby when a cupboard door is opened a light within it is lit up can be done by means of a normally off switch specially provided for this purpose. These switches have a protruding knob which is arranged that it may be pressed by the back of the cupboard door as the door closes, and this action holds the switch open. As the door is opened the knob springs out and closes the circuit. These switches can be inserted into the frame of the door in a specially made hole, or they may be mounted inside the frame with a wooden block on the door carefully arranged to press on the knob and open the switch as the door closes.

Connections to gas appliances and oil-fired central heating plant

There may be a need to provide a supply for the equipment that is associated with central-heating equipment, either for lighting up the flame when a thermostat determines when it should be lit, or for increasing water flow in a central-heating water system circuit.

The supply to such devices may be made by a normal socket-outlet, but the cable used must be of the butyl heat-resistant type.

In certain larger installations, it may be necessary to use flameproof switches and other fittings in the room where the gas- or oil-heated boiler is situated. Consultation with the manufacturers may be necessary.

Connections to underfloor electric heating

For bungalows especially, underfloor electric heating is frequently employed. This form of heating may be carried out in several ways, but it often consists of insulated electric wires buried in the floor and covered by a layer of concrete. Some systems employ plastic-coated mineral-insulated copper-sheathed cable, which may be safely operated at high temperatures, others employ special heating wires simply immersed in the concrete, while what is probably the best system uses a conduit buried in the cement of the floor and withdrawable heating cables.

For a room of 5 m by 4 m a floor heating loading of about 3 kW is necessary, and this is better supplied by means of a separate feed run directly back to the consumer unit. This is because a thermostat is needed in the circuit, to prevent undue consumption of current, and this would mean a wiring complication if the floor heating was supplied from the normal ring circuit.

In any case, such supplies should always be taken at off-peak periods, so that a separate feed is automatically needed.

The actual designing and installation of a floor-heating system requires specialist knowledge, and in any case must be carried out over a considerable period in close collaboration with the builder. Such a system can very rarely be installed in any but new buildings.

Warm-air central heating and air conditioning

Some houses are equipped with ducted air systems, so that air is warmed up at a central point (or at several points) by electric heaters, with fans to control the flow of air.

In providing supplies to these heaters, the only point the electrician has to observe is to ensure that the cables are of adequate size (since some heating elements exceed 3 kW in capacity) and that the necessary thermostat connections, as required by the makers, are incorporated into the wiring layout.

In many houses complete air conditioning is employed. This means that the air taken into the sealed, double-glazed and fully insulated interior of the house is cleaned, humidified if necessary, and heated in winter or cooled in summer, the air-conditioning plant being provided with both heating and refrigerating units.

Some of these air conditioners have complex thermostat and humidity-controlling devices, which need extensive wiring runs; and since the load on many of them is quite high – 12 kW is common for an ordinary three-bedroomed house – obviously special large current-carrying mains cables and appropriate fuses are needed.

Dimmers

To dim a domestic lighting circuit thyristors are widely used. The thyristor blocks the current for part of each cycle of alternation, the amount of blockage being controlled by the application of a control voltage to one of the electrodes of the thyristor by means of a variable resistor. Such a device is inserted in the lighting circuit and allows for full control, without the problem of heat loss, and in relatively small bulk.

Another commonly used method in the theatres and dance halls is to insert a variable resistance between the mains and the lamp circuits to be dimmed.

This resistance obviously carries the whole current of the lighting system, and so will develop heat. Moreover, until recently most resistance dimmers of this type were fairly bulky, and could not therefore be easily incorporated into the lighting circuits of, say, an ordinary lounge or dining room.

Nowadays, devices have been developed of a size not very much larger than a socket-outlet, which can be safely buried in the wall, and are arranged so that the heat is safely dissipated.

Delay switches

In buildings where there are a number of flats with access by stairways instead of lifts, delay switches are sometimes fitted, to allow for the lights to be switched on when entering the building at basement level, and for them to remain on for a given period of perhaps five minutes. They then switch themselves off to avoid current waste in lighting up the staircases and passages during the night, when no light is needed.

These delay switches take several forms. In some the pressure on the switch winds up a small spring, which then unwinds over a period and switches off the circuit. In another form a pneumatic piston is provided within the switch, which is lowered into its cylinder against the air pressure by the operation of switching on, and then as the air gradually escapes through a controlled orifice, the piston rises with the help of a spring and switches off the lamp. These devices can be purchased from most switch manufacturers.

Security devices

There are several patent burglar-alarm systems on the market, but many of them have the disadvantage that being widely advertised the burglars themselves are fully familiar with their requirements. The best system is undoubtedly one evolved specially for the installation concerned. Obviously

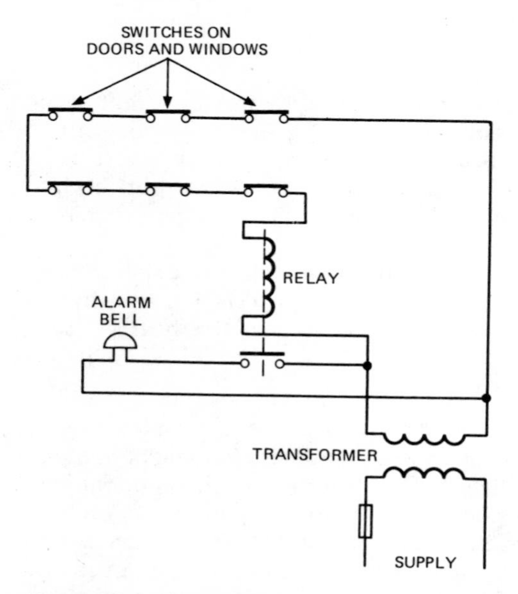

Figure 8.5. A burglar alarm circuit, using a relay, so that a broken wire or opened switch will cause an alarm bell to sound

all windows and doors need some kind of protection, and small switches will have to be installed on every door and window to operate if the door or window is opened. A simple 'closed' circuit system may be wired so that even if the burglar sees the wiring and cuts it, the alarm still operates. This system is arranged as shown in *Figure* 8.5. Each door or window is fitted with a small switch so that if entry is attempted, the switch contacts will open. This will cause the relay to release and close a bell circuit, giving an alarm. If wires are cut or break, the alarm will sound, thus the system is self-monitoring, against open circuits.

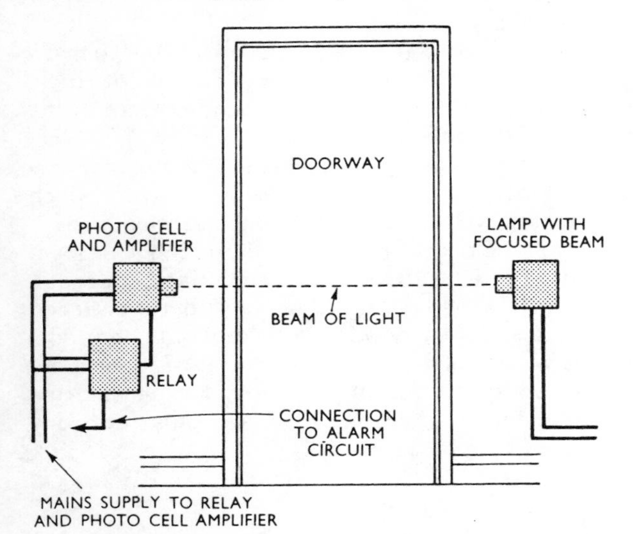

Figure 8.6. The photoelectric cell and the wiring necessary

Photoelectric cell circuits may be used also for security systems (*Figure* 8.6). A photoelectric cell is a device which conducts current when light is thrown upon it, or conversely, stops conducting when light is withdrawn. A beam from a lamp is focused by lenses across an aperture such as a doorway which is to be protected, and falls on to a photoelectric cell, the current from which is amplified by a mains or battery operated amplifier circuit, and this current operates a relay which is normally held open. If the beam of light is interrupted by the passage of a person or other object, the relay falls off and the alarm circuit is connected.

A variant sometimes used is to employ an infra-red beam that cannot be seen by the human eye, and this can operate a special type of electric cell.

The wiring for these devices follows the standard rules as regards installation, but the actual connections are given by the makers of this type of equipment, since each type differs.

Baby-alarm circuits

One facility often provided is the baby-alarm circuit, so that a microphone can be placed beside the baby's cot in an upstairs room, and if the child cries it can be heard in the sitting room, dining room or kitchen.

The microphone should be a moving-coil type, commonly employed for tape recording, since the crystal type (which are the cheapest) will not work with long extension leads.

The microphone points may well be flush VHF types and they should be wired with television-type cable back to the central amplifier which will have running from it switched loudspeaker lines (ordinary twin plastic cable can be used).

The same wiring rules apply as in other cases, which are that the wires are best run in plastic conduit beneath the plaster, but in any case must not have any contact with any type of circuit fed from the mains.

Emergency lighting

Emergency lighting can be provided in ordinary domestic premises, with moderate cost, by using a trickle charger suitable for a 12 V car battery, the battery itself, and a relay (see *Figure* 8.7).

The system requires a relay which will remain permanently energised as long as its coil is connected to the 240 V mains circuit. This relay should be properly protected on the mains side with a 2 A fuse, and should be enclosed to prevent damage.

The contacts, which remain permanently open when the relay is energised, may be connected to the battery and to a 12V lighting circuit run as required in the house. This wiring must be kept quite separate from any other wiring. IEE Reg. 525–2 to 7. Twin plastic cable run in plastic conduit will be

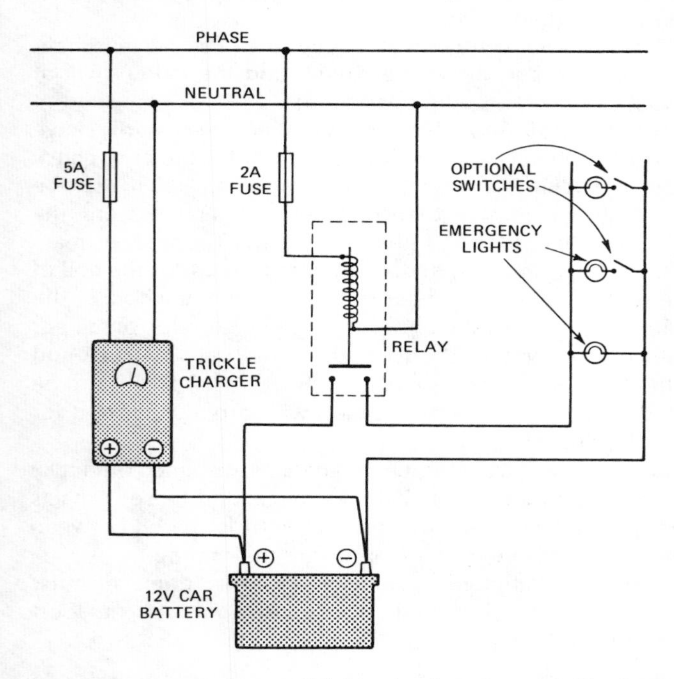

Figure 8.7. An emergency lighting circuit, using a trickle-charged 12V automobile-type battery as the main source of supply

quite satisfactory, but surface wiring neatly cleated is of course permissible.

Obviously emergency lighting is not required everywhere, but one 36W car headlamp bulb in the sitting room, another in the kitchen and perhaps a 6W type on the landing and a further 6W bulb in a lavatory will be sufficient.

When the relay drops off through mains failure, all these lamps will be lit. As mains failure is usually of short duration, it may not matter if the lamps remain alight, even if all of them are not needed. In any case, the battery benefits by being discharged on occasion. If no failures occur for long periods, the battery may deteriorate.

When the mains supply is restored, the relay will automatically pick up and break the circuit, and the battery will at once begin to charge up once more.

The type of relay obtained may not necessarily have contacts that will carry the whole current for the emergency lighting circuit if all the lamps are in use, and it may be necessary to employ a second relay, operated from the battery, with contacts of greater capacity. The first, or main failure relay, simply operating a battery circuit for the coil of the second relay, whose contacts will carry the whole of the main current. This depends on the type of relay obtained. The current for a 36W car headlamp bulb at 12V is 3A and thus if two of these are in use plus two 6W lamps, 7A will be needed, and not all relay contacts will carry this current for any length of time.

Some relays have a number of contacts, designed originally for closing a number of circuits at once, and these contacts can be paralleled up so that the current is shared between them and they are all working within their rating.

Emergency lighting for premises other than domestic must comply with BS 5266 and the requirements of the local licensing authority.

Electricity in the greenhouse and the garden

Electricity may be used in a greenhouse for heating, for seed propagation, for raising cuttings and for soil sterilisation, and many other purposes. It may also be used in the garden to drive a lawnmower or hedgecutter.

Great care should be taken when installing socket-outlets in greenhouses where damp and humid conditions often prevail, to ensure that all fittings are of the weatherproof type, and to ensure proper earthing, since an electric shock under these conditions would be especially dangerous.

Conduit systems, mineral-insulated copper-sheathed or PVC armoured cables, with metallic fittings are the best for this purpose.

It is very desirable that such installations should be protected by residual-current-operated earth-leakage circuit breakers. This protection, operating at not more than 30 mA, is compulsory for socket-outlets which may feed appliances outdoors, such as lawnmowers or hedgecutters – see Chapter 3.

The ELCB can be connected so as to protect the whole installation or alternatively socket-outlets incorporating a residual-current operated ELCB are available.

Index